LOGARITHMIC AMPLIFICATION

The Artech House Radar Library

LOGARITHMIC AMPLIFICATION

With Application to Radar and EW

Richard Smith Hughes

International Standard Book Number: 0-89006-182-3
Library of Congress Catalog Card Number: 85-073632

THIS WORK IS RESPECTFULLY
DEDICATED TO THE MEMORY
OF MY PARENTS

FRANCES SMITH HUGHES[†]
JOHN RICHARD HUGHES[†]

PREFACE

Logarithmic amplifiers are an indispensable part of many modern radar and electronic warfare systems. They come in two basic versions, logarithmic intermediate frequency amplifiers and detector-logarithmic video amplifiers, and both have distinct strong and weak points dependent on particular system requirements. The detector-logarithmic video amplifier has an input frequency range up to 20 GHz, while present day logarithmic intermediate frequencies are limited to 2 GHz or so. The input dynamic range for logarithmic intermediate frequency amplifiers is much larger than their detector-logarithmic video amplifier cousins. The logarithmic conformity is generally better for detector-logarithmic video amplifiers than logarithmic intermediate frequencies; however, the logarithmic intermediate frequency has generally superior pulse characteristics. Thus, the choice of what logarithmic amplifier type to use for a given system may well be uniquely tied to that system's mission and wanted characteristics. This book, although design oriented, presents the theoretical and practical characteristics of detector-logarithmic video amplifiers and logarithmic intermediate frequency amplifiers with the strengths and weaknesses of each, to aid the radar and electronic warfare system engineer in determining the best choice of logarithmic amplifier for their system.

The book is comprised of six chapters and eight appendices that take the reader from the basic logarithmic transfer function to continuous wave rejection (effect of continuous wave on the pulse response). Chapter 1 presents the basic logarithmic transfer function. Logarithmic video amplifiers are presented in Chapter 2 with logarithmic intermediate frequencies presented in Chapter 3. Chapter 4 presents the quest for direct current-coupled detector-logarithmic video amplifiers (a most difficult design problem which is far less complicated with logarithmic intermediate frequencies). Chapter 5 discusses the pertinent parameters the designer and system engineer must consider before the design (or purchase) of a logarithmic amplifier is made. Chapter 6 presents several simple applications using logarithmic amplifiers. An extensive bibliography on logarithmic amplification and kindred subjects is found at the end of the book.

The author would like to extend his appreciation to John Daugherty and Paul Hilliard, who during the past 15 years constructed and tested all the circuits presented. Cliff Hauser and Kris Kauffman have critiqued my thoughts over the years, and I am deeply appreciative of their constructive discussions. The present work was reviewed by Cliff Hauser, Kris Kauffman, and Bob Sutton and their comments and suggestions were greatly appreciated. I wish to give special thanks to my Naval Weapons Center (NWC) Editors, Doris Quick and Alma Barber, who have suffered through my hieroglyphs to produce readable documents for more years then we care to remember. My deepest thanks and appreciations go to the management of NWC (past and present), for creating the practical and academic atmosphere necessary to produce this work. Last, but by no means least, my thanks to my wife, Janet, for her undying support.

A note on security: As an engineer employed by the U.S. Government, I am acutely aware of the possibilities of inadvertently and unwittingly compromising Navy projects. To remove this risk, frequencies are talked of only in generalities. Detector-logarithmic videos are useful for radio frequencies up to 20 GHz. The maximum frequency for logarithmic intermediate frequency amplifiers is currently around 2 GHz. Both of these statements are contained in vendors' catalogs. The detector–logarithmic video amplifier designs presented are for frequencies where commercial logarithmic intermediate frequencies are easily obtained so comparisons between the two may be made. The chapter on selected applications is much shorter and less descriptive than desired, as any configuration or discussion that I felt could, in any way, compromise the work at NWC in which we are engaged was left out completely.

CONTENTS

Chapter 1

THE BASIC LOGARITHMIC TRANSFER FUNCTION

Introduction

Logarithmic amplification is an indispensable tool in many radar and electronic warfare (EW) receivers. Linear amplification (automatic gain control (AGC)) has a limited useful instantaneous input dynamic range (10 decibels (dB) or so) before saturation occurs. The logarithmic amplifier compresses a much larger input dynamic range into a small output dynamic range, and logarithmic amplifiers with an instantaneous input dynamic range in excess of 80 dB are not uncommon.

This chapter presents a classic use for logarithmic amplifiers (a simple mono-pulse direction finder (DF)), the theoretical logarithmic transfer function, and some practical deviations from the ideal.

Why Logarithmic Amplification?

Figure 1-1 illustrates a basic monopulse DF [1, 2].* Two antennas have their axes tilted and provide a gain pattern as a function of angle (θ) [3]. Consider a signal in the position shown.** The input signal to the two receivers (mixing is omitted for simplicity) is a function of transmitted power and frequency as well as a function of the antenna gain, $G(\theta)$. Assume that the two antennas are physically close together such that the signal at the faces of the two are equal. Letting I represent this signal, then

$$e_{o|A} = G_A(\theta)I \qquad (1\text{-}1)$$

---

*Numbers in brackets [ ] refer to references at the end of each chapter. A comprehensive bibliography is presented at the end of the book.

**These antenna patterns are highly simplified, and the antenna gains are normalized.

1

FIGURE 1-1. Basic Monopulse Direction Finder.

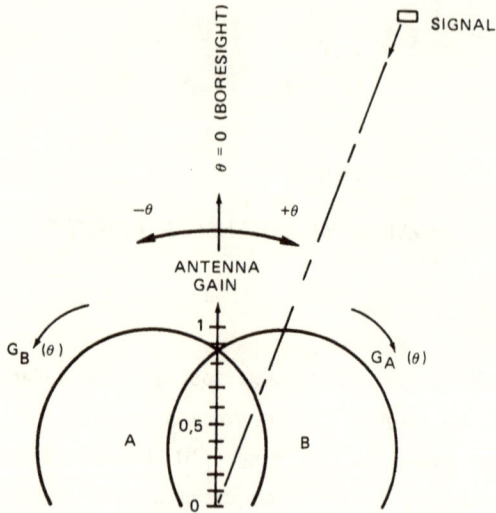

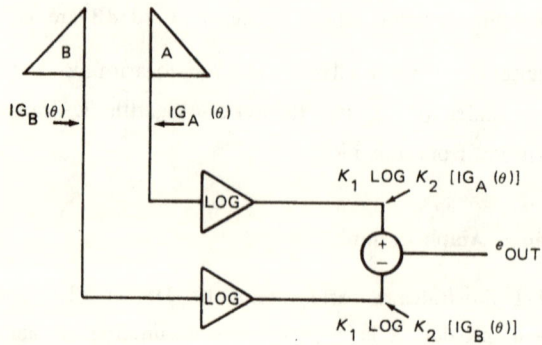

$$e_{OUT} = K_1 \ LOG \ K_2 \ IG_A \ (\theta) - K_1 \ LOG \ K_2 \ IG_B \ (\theta)$$

or

$$e_{OUT} = K_1 \ LOG \ \frac{G_A \ (\theta)}{G_B \ (\theta)}$$

and

$$e_{o|B} = G_B(\theta)I \qquad (1\text{-}2)$$

where $G_A(\theta)$ and $G_B(\theta)$ are the antenna gains as a function of angle (θ). Obviously, these two signals are a function of both $G(\theta)$ and the signal at the antennas' face I. I can change from microvolts (μV) to volts (V) in a short time span (depending on range, frequency, antenna lobe pattern, etc.), and AGC has been used to normalize this widely varying input dynamic range [4, 5]. The use of logarithmic amplifiers removes the need for the AGC amplifiers (and associated circuitry) and also makes use of the unique properties of logarithmic arithmetic.

The general transfer function for a logarithmic amplifier may be given as

$$e_{out} = K_1 \log K_2 e_{in} \qquad (1\text{-}3)$$

where K_1 and K_2 are constants associated with the logarithmic amplifier. The outputs of the two logarithmic amplifiers in Figure 1-1 may now be given as (substituting Equation 1-1 and 1-2 into Equation 1-3)

$$e_{out|A} = K_1 \log K_2 \, IG_A(\theta) \qquad (1\text{-}4)$$

$$e_{out|B} = K_1 \log K_2 \, IG_B(\theta) \qquad (1\text{-}5)$$

Taking the difference of the two signals

$$\Delta \equiv e_{out|A} - e_{out|B} \qquad (1\text{-}6)$$

or

$$\Delta = K_1 \log K_2 \, IG_A(\theta) - K_1 \log K_2 \, IG_B(\theta) \qquad (1\text{-}7)$$

or, since $\log A - \log B = \log A/B$, Equation 1-7 may be given as

$$\Delta = K_1 \log \left(\frac{K_2 \, IG_A(\theta)}{K_2 \, IG_B(\theta)} \right) \qquad (1\text{-}8)$$

or

$$\Delta = K_1 \log \frac{G_A(\theta)}{G_B(\theta)} \qquad (1\text{-}9)$$

3

The result given in Equation 1-9 is most important because the resultant signal, Δ, represents the angle of arrival, both in terms of magnitude and sign, independent of the signal input intensity. This angle-of-arrival signal, Δ, is valid each time the signal is present; thus only one pulse (monopulse) is needed to determine the angle of arrival.

The ideal and practical logarithmic transfer function is discussed before presenting the methods of obtaining a logarithmic transfer function.

The Ideal Logarithmic Transfer Function

Equation 1-3 defined the ideal logarithmic amplifier transfer function. In practice, we never quite obtain this ideal; however, it is most worthwhile to spend a little time with Equation 1-3. The discussions of this and the next section assume a voltage input to the logarithmic amplifier; however, the marriage of a radio frequency (RF) (or intermediate frequency (IF)) detector to the logarithmic video amplifier is mathematically straightforward and is discussed in the following chapters.

Equation 1-3 may be written as

$$e_{out} = K_1 \log e_{in} + K_1 \log K_2 \tag{1-10}$$

and is shown graphically in Figure 1-2 for three sets of values of K_1 and K_2 [6].

It is of value to solve this equation for K_2 when $e_{out} = 0$

$$K_1 \log e_{in} = -K_1 \log K_2 \tag{1-11}$$

or

$$K_2 = \frac{1}{e_{in}} \text{ (for } e_o = 0) \tag{1-12}$$

and K_2 may be visualized as determining the start of logarithmic action (Figure 1-2a). The larger K_2, the smaller e_{in} for $e_{out} = 0$. Changing K_1 changes the degree of output compression (increasing K_1 decreases the compression (see Figure 1-2a)).

4

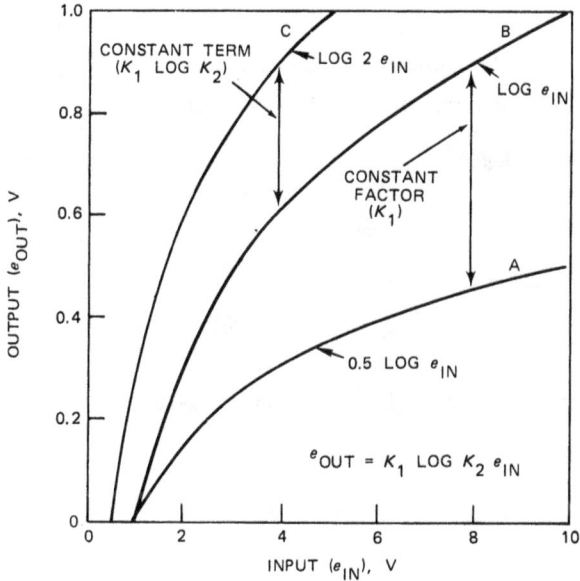

A: $K_1 = 0.5$, $K_2 = 1$
B: $K_1 = K_2 = 1$
C: $K_1 = 1$, $K_2 = 2$

(a) e_{in} linear scale.

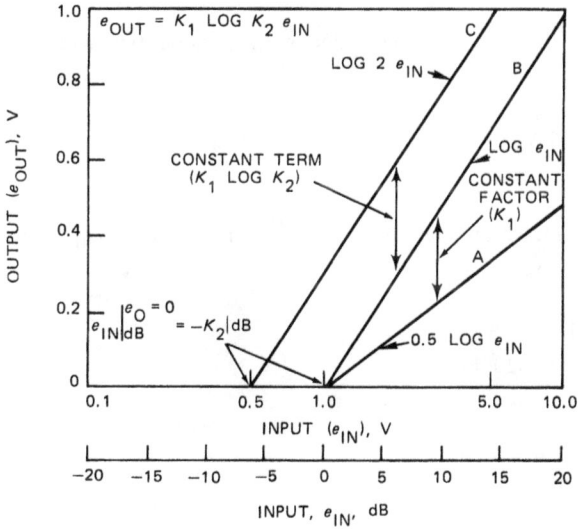

(b) e_{in} logarithmic and decibel scale.

FIGURE 1-2. Characteristics of Equation 1-3.

5

However, this is best illustrated by Figure 1-2b, which has a logarithmic x-axis. As can be seen, the output is a straight line. We will go one step further and define the x-axis in terms of decibels.* Now, since

$$e_{in}|_{dB} = 20 \log e_{in} \tag{1-13}$$

e_{in} can be solved for

$$e_{in} = 10^{\frac{e_{in}|_{dB}}{20}} \tag{1-14}$$

Substituting Equation 1-14 into Equation 1-10

$$e_{out} = K_1 \log 10^{\frac{e_{in}|_{dB}}{20}} + K_1 \log K_2 \tag{1-15}$$

or

$$e_{out} = \frac{K_1}{20} e_{in}|_{dB} + K_1 \log K_2 \tag{1-16}$$

Equation 1-16 is a straight line with a slope (we will call this the logarithmic slope (LS)) of

$$LS = \frac{K_1}{20} \frac{V}{dB} \tag{1-17}$$

The start of logarithmic action $e_{in}|_{dB}^{e_o=0}$ ($e_{in}|_{dB}$ for $e_{out} = 0$) may now be found from Equation 1-16 as

$$e_{in}|_{dB}^{e_o=0} = -20 \log K_2 \tag{1-18}$$

*In the voltage domain, decibel is used to define voltages with respect to 1 V or decibel = $20 \log e_{in} = e_{in}|_{dB}$.

6

or

$$e_{in}\Big|_{dB}^{e_o=0} = -K_2\Big|_{dB} \text{ (for } e_{out} = 0) \tag{1-19}$$

Thus, logarithmic action starts at $-K_2\big|_{dB}$, and the output increases $K_1/20$ V for each decibel increase in the input (Figure 1-2b). Equation 1-16 may be given as

$$e_{out} = \frac{K_1}{20} e_{in}\big|_{dB} + \frac{K_1}{20} K_2\big|_{dB} \tag{1-20}$$

or, substituting Equations 1-17 and 1-19 into Equation 1-20

$$e_{out} = LS \left(e_{in}\big|_{dB} - e_{in}\Big|_{dB}^{e_o=0}\right) \tag{1-21}$$

The general equations for Figure 1-2b may now be given, for curve A as

$$K_1 = 0.5, \; K_2 = 1 \qquad e_{out} = 25 \; \frac{mV}{dB} \left(e_{in}\big|_{dB}\right) \tag{1-22}$$

or the output increases 25 millivolts (mV) for each decibel increase above the minimum input $e_{in}\Big|_{dB}^{e_o=0} = 20 \log K_1$ (0 dB); for curve B as

$$K_1 = 1, \; K_2 = 1 \qquad e_{out} = 50 \; \frac{mV}{dB} \left(e_{in}\big|_{dB}\right) \tag{1-23}$$

where

$$e_{in}\Big|_{dB}^{e_o=0} = 0 \text{ and } LS = 50 \text{ mV/dB, and for curve C as}$$

$$K_1 = 1, \; K_2 = 2 \qquad e_{out} = 50 \; \frac{mV}{dB} \left(e_{in}\big|_{dB} + 6\right) \tag{1-24}$$

where

$$e_{in}\Big|_{dB}^{e_o=0} = -6 \text{ dB and LS} = 50 \text{ mV/dB}$$

In practice, the designer can control where logarithmic action starts, and K_1, which determines the logarithmic slope (LS).

The Practical Logarithmic Transfer Function

The ideal transfer function of Equation 1-20 (see Figure 1-2b) is what this book is all about; however, modern logarithmic amplifiers, suitable for radar and EW applications, are not truly logarithmic. They approximate a logarithmic curve but never *exactly* exhibit the characteristics of Equation 1-20.

Figure 1-3 illustrates the transfer function for a practical logarithmic amplifier. The actual output varies around an ideal logarithmic transfer function with a logarithmic error, $\pm\epsilon(dB)$. This error (logarithmic conformity) is one of the design parameters we will discuss in the next two chapters. However, errors of less than ±0.5 dB are fairly easily obtained. As discussed earlier, the ideal transfer function crosses the input axes at $-K_2(dB)$. The transfer function for low–level signals is usually linear for output signal-to-noise (S/N) reasons.

The output S/N ratio for a linear amplifier increases 1 dB for each decibel increase in input signal level. Due to the compression of a logarithmic amplifier, the output increases less than 1 dB for each decibel increase in the input signal; thus the output S/N increases less for a given increased input to a logarithmic amplifier than for a linear amplifier. This is generally a concern if the logarithmic amplifier is to drive a signal threshold.*

---

*The sensitivity for a receiver depends upon the S/N ratio at the input to signal threshold. Reference 7 gives a general discussion.

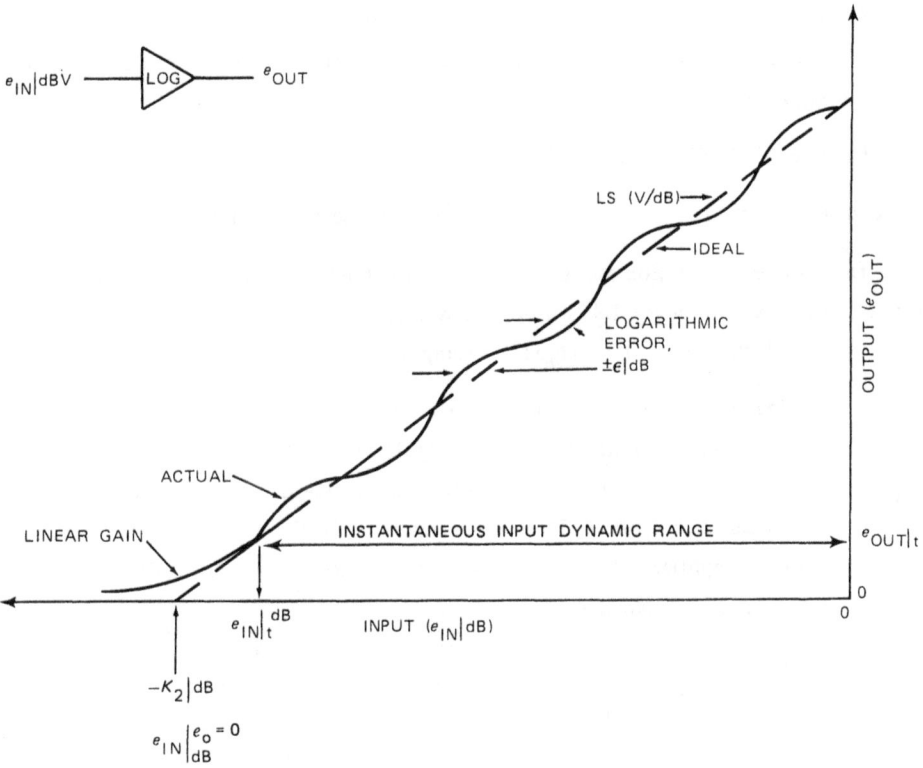

FIGURE 1-3. Practical Logarithmic Transfer Characteristic.

Thus, the low–level transfer function of practical logarithmic amplifiers is a linear function of the input signal until the desired linear–logarithmic threshold is reached

$$e_{in}\Big|_t^{dB} = \text{linear–logarithmic transition input}$$

$$e_{out}\Big|_t = \text{output for } e_{in}\Big|_t^{dB}$$

The logarithmic amplifier designer controls not only where the logarithmic action starts but the LS as well. There are two practical limits on LS (thus K_1, Equation 1-17). Available power supply voltage and power dissipation set an upper limit on LS, and any post–logarithmic amplifier circuitry sets the lower limit.

9

Suppose only 6 V are available at the output stage of a logarithmic video amplifier, and a 70–dB input dynamic range is wanted. The maximum LS may be approximated as

$$LS \cong 6 \text{ V}/70 \text{ dB} = 85.7 \text{ mV/dB} \tag{1-25}$$

Increasing the LS may also deteriorate the output rise time (to be discussed later).

The primary limitation on LS is not usually how large, but rather how small it can be. The lower limit for LS is usually determined by any post–logarithmic amplifier circuitry, as illustrated in the following example:

A logarithmic amplifier is to drive a follow–hold amplifier* that has a direct current (DC) offset, drift, charge transfer,** etc., of 50 mV. If the logarithmic amplifier has a 10–mV/dB LS, this 50–mV offset represents a 5–dB error referred to the input of the logarithmic amplifier. If the error is not to exceed, say, 0.25 dB, the LS must be no smaller than

$$LS = \frac{50 \text{ mV}}{0.25 \text{ dB}} = 200 \frac{\text{mV}}{\text{dB}} \tag{1-26}$$

Figures 1–4 and 1–5 illustrate the transfer characteristics of a detector–logarithmic video amplifier that is discussed in the next chapter. The design specifications for this amplifier are

Input dynamic range $\geqslant$40 dB requiring LS = 17.4 mV/dB

Logarithmic error (or logarithmic conformity), $\epsilon|_{dB} < \pm 0.5$ dB

Linear–logarithmic transition input power, $P_{in}\big|_t^{dBm} \cong -40$ dBm***

---

*Sometimes called track–hold amplifier.

**The output of a follow–hold amplifier will follow (track) the input until a hold command stores the voltage (usually on a capacitor). There is some charge transfer from the hold command to the holding capacitor.

***dBm = decibel referred to 1 milliwatt.

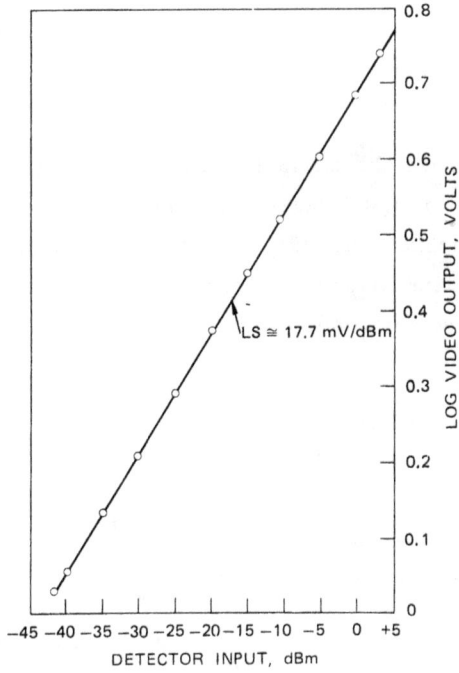

FIGURE 1-4. Detector-Logarithmic Video Amplifier (see Figure 2-36).

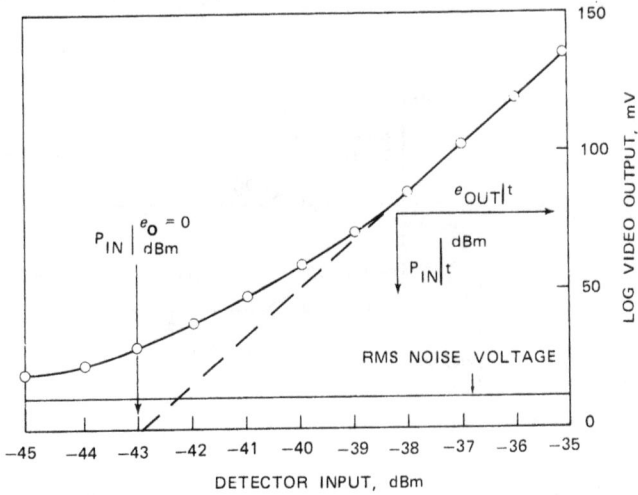

FIGURE 1-5. Detector-Logarithmic Video Amplifier (see Figure 2-36).

11

The characteristics, illustrated in Figures 1-4 and 1-5, easily meet the design requirements. The importance of this is that the designer has tight design control to meet specific requirements.

We have just covered the practical logarithmic transfer function. The next two chapters present the two most important logarithmic amplifier techniques: (1) detector logarithmic video amplifier and (2) logarithmic IF amplifier. However, it is of interest to review several receiver techniques before proceeding. Figure 1-6 is a summary of various receiver configurations (the use of limiting IF or RF amplifiers will not be discussed).

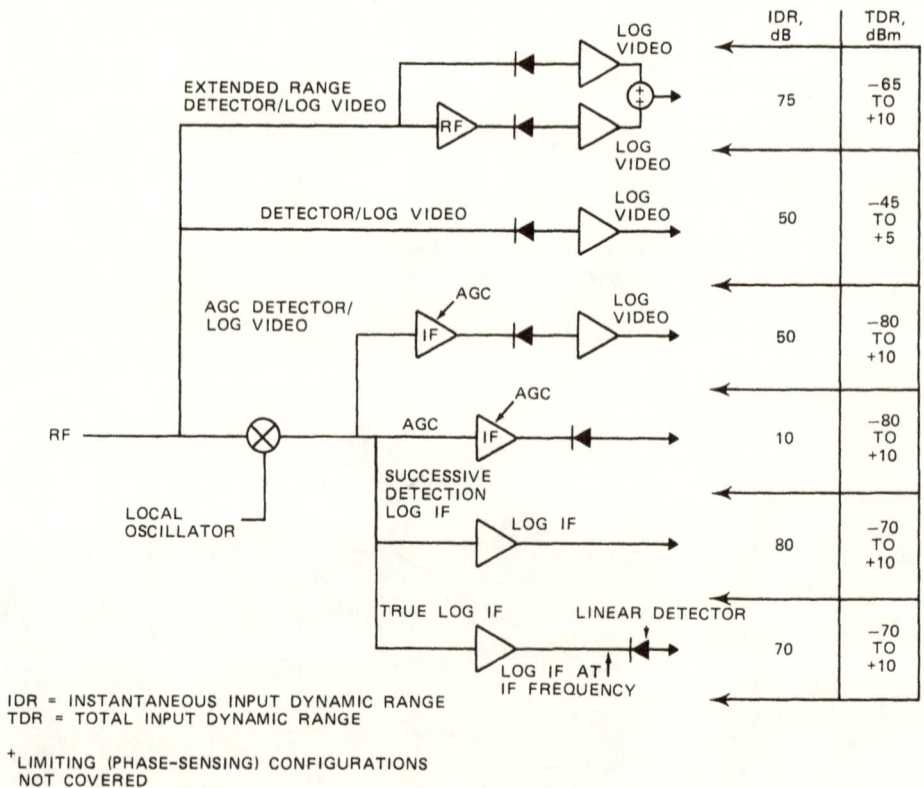

IDR = INSTANTANEOUS INPUT DYNAMIC RANGE
TDR = TOTAL INPUT DYNAMIC RANGE

+LIMITING (PHASE-SENSING) CONFIGURATIONS
NOT COVERED

FIGURE 1-6. Basic Radar/Electronic Warfare Receiver Techniques.

12

The most common receiver uses AGC and has a limited instantaneous input dynamic range of some 10 dB (20 dB if a linear detector is used) and a wide total dynamic range (90 dB). If a wide instantaneous input dynamic range at IF or RF is wanted, either logarithmic IF amplifiers or a detector–logarithmic video amplifier (in conjunction with AGC to provide extended total dynamic range) may be used. Logarithmic IF amplifiers have a wide instantaneous dynamic range (80 dB); however, their IF is limited to below 2 gigahertz (GHz). The bandwidth for logarithmic IF amplifiers increases as the IF increases with a maximum around 400 to 700 megahertz (MHz). The detector–logarithmic video amplifier has a smaller instantaneous input dynamic range (50 dB or so); however, frequencies in excess of 18 GHz with useful bandwidths in excess of 15 GHz may be obtained. The instantaneous dynamic range can be increased by the use of RF amplifiers (and associated RF coupling, etc.) at the expense of RF and bandwidth.

Choosing which type of logarithmic technique to use depends upon the characteristics a given system requires. The various trade–offs will be discussed as we proceed.

References

1. Lipsky, S. E., "Log Amps Improve Wideband Direction Finding," *Microwaves*, May, 1973.

2. - - - - - -. "Find the Emitter Fast With Monopulse Methods," *Microwaves*, May, 1978.

3. Cohen, W. and C. M. Steinmetz, "Amplitude– and Phase–Sensing Monopulse System Parameters," *Microwave Journal*, October 1959 and November 1959.

4. Naval Weapons Center. *Automatic Gain Control; A Practical Approach to Its Analysis and Design*, by R. S. Hughes. China Lake, Calif., NWC, August 1977. (NWC TP 5948.)

5. Hughes, R. S., "Design Automatic Gain Control Loops the Easy Way," *EDN*, October 5, 1978.

6. ------. *Logarithmic Video Amplifiers*, Dedham: Artech House, 1971.

7. Naval Weapons Center. *Determining Maximum Sensitivity and Optimum Maximum Gain for Detector-Video Amplifiers With RF Preamplification*, by R. S. Hughes. China Lake, Calif., NWC, February 1985. (NWC TM 5357.)

Chapter 2

LOGARITHMIC VIDEO AMPLIFIERS

Detector-logarithmic video amplifiers are useful for input frequencies below 100 MHz to frequencies in excess of 18 GHz and bandwidths greater than 10 GHz. They exhibit a limited instantaneous input dynamic range (–45 to +15 dBm if Schottky detectors are used and –40 to +5 dBm for tunnel diode detectors) with comparison to logarithmic IF amplifiers (–80 to +10 dBm).

This chapter presents a review of several logarithmic video techniques to stimulate the reader's interest. A design procedure for the parallel summation detector-logarithmic video is then presented (this is by far the most common method currently in use to obtain radar and EW quality detector-logarithmic video amplifiers). This chapter closes with a series linear-limiting (lin–limit) technique that offers distinct advantages if narrow pulses (sub 50 nanoseconds (nsec)) are to be handled.

Review of Various Logarithmic Video Amplifier Techniques

Diode, or transconductance, feedback (Figure 2-1) is a common method used to obtain an excellent logarithmic response [1,2]; however, this technique has found limited, if any, use in systems requiring high quality, large input dynamic range, fast rise time logarithmic video amplification [3]. A more powerful logarithmic technique is the approximation of a logarithmic function by the summation of straight, or curved, line segments. These techniques are inherently useful, since the logarithmic elements can be designed to compensate for any input nonlinearities (i.e., it is simple to compensate for crystal saturation at high input levels).

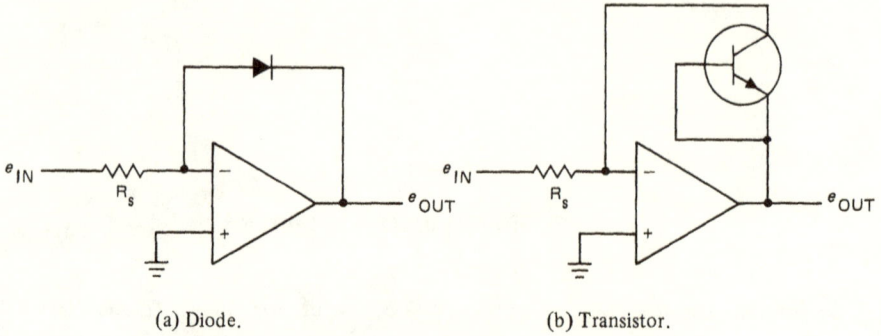

(a) Diode.   (b) Transistor.

FIGURE 2-1. Transconductance Feedback Logarithmic Amplifiers.

The general logarithmic relationship is given by Equation 1-3 as

$$e_{out} = K_1 \log K_2 e_{in} \tag{2-1}$$

This characteristic may be approximated by summing outputs that have a common difference, Δ, that result from a series of inputs having a common ratio R. This can be demonstrated as follows. Let the input voltages $e_{in_0}, e_{in_1}, e_{in_2}, \cdots e_{in_n}$, have a common ratio, R, such that

$$\frac{e_{in_n}}{e_{in_{n-1}}} = \ldots = \frac{e_{in_2}}{e_{in_1}} = \frac{e_{in_1}}{e_{in_0}} = R \tag{2-2}$$

Thus

$$e_{in_n} = e_{in_0} R^n \tag{2-3}$$

Substituting Equation 2-3 into 2-1 gives

$$e_{out_n} = K_1 \log \left[K_2 \left(e_{in_0} R^n \right) \right] \tag{2-4}$$

or

$$e_{out_n} = K_1 \log K_2 e_{in_0} + nK_1 \log R \tag{2-5}$$

However, since

$$e_{out_0} = K_1 \log K_2 e_{in_0} \tag{2-6}$$

16

Equation 2-5 reduces to

$$e_{out_n} = e_{out_o} + nK_1 \log R \qquad (2\text{-}7)$$

Equation 2-7 indicates that a logarithmic relationship specifies a series of inputs having a common ratio, R, corresponding to a series of outputs having a common difference, Δ, where

$$\Delta = K_1 \log R \qquad (2\text{-}8)$$

(Insight into the significance of Equation 2-7 is given shortly.)

One method of generating the proper series is by summing, in parallel, cascaded lin-limit amplifier stages [3]. Figures 2-2 and 2-3 illustrate a single lin-limit characteristic and a series of cascaded stages driving a summing amplifier, respectively.

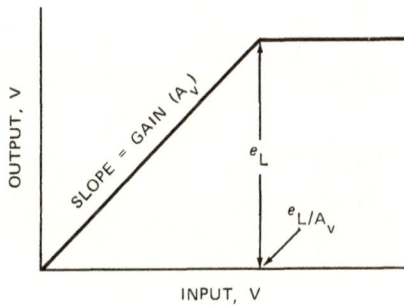

FIGURE 2-2. Linear-Limiting
Input-Output Characteristics.

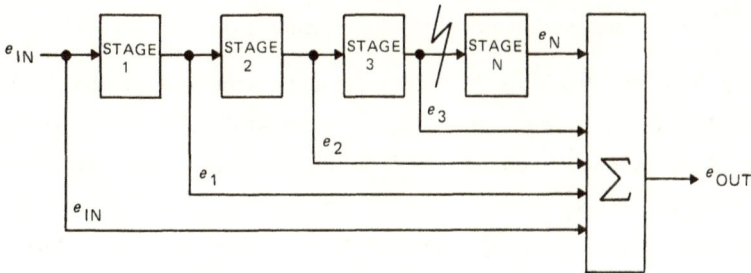

FIGURE 2-3. Basic Parallel Linear-Limiting Logarithmic Amplifier.

17

The summed output for Figure 2-3 is

$$e_{out} = e_{in} + e_1 + e_2 + e_3 + \ldots e_n \tag{2-9}$$

Assume that the input is slowly increased until all stages have limited (let n = 4),

$$\left. e_{out} \right|_{\substack{e_L/A_V^4 \\ 4\text{limit}}} = e_L + \frac{e_L}{A_V} + \frac{e_L}{A_V^2} + \frac{e_L}{A_V^3} + \frac{e_L}{A_V^4}$$

$$\left. e_{out} \right|_{\substack{e_L/A_V^3 \\ 3\text{limit}}} = 2e_L + \frac{e_L}{A_V} + \frac{e_L}{A_V^2} + \frac{e_L}{A_V^3}$$

$$\left. e_{out} \right|_{\substack{e_L/A_V^2 \\ 2\text{limit}}} = 3e_L + \frac{e_L}{A_V} + \frac{e_L}{A_V^2} \tag{2-10}$$

$$\left. e_{out} \right|_{\substack{e_L/A_V \\ 1\text{limit}}} = 4e_L + \frac{e_L}{A_V}$$

where

$$\left. e_{out} \right|_{\substack{e_L/A_V^n \\ n\text{limit}}} = \text{output for stage n limited (input is } e_L/A_V^n)$$

$$e_L = \text{single-stage limiting value (Figure 2-3)}$$

If $A_V \gg 1$, the terms to the right of the e_L term in Equation 2-10 may be neglected. As will be shown shortly, A_V must be small (3 to 6) to give a reasonable logarithmic response. Equation 2-10 fulfills the requirement that by increasing the input by a constant ratio (the gain of the amplifier (A_V)), the output increases by a constant term, e_L, (or the limiting value of the amplifier). Thus, the configuration of Figure 2-3 has a logarithmic response at the breakpoints (that point which the output of a stage limits). The composite response is illustrated in Figure 2-4. As will be seen, the output has a true logarithmic response at the lin-limit transitions (e_L/A_V^n, etc.); however, an error does exist between the transitions and is a function of the ratio (the amplifier gain). Figure 2-5 [4,5] illustrates the error, referred to the input, as a function of the gain, A_V in dB ($A_V|_{dB}$). It will be seen that the error decreases for decreasing gains. This is reasonable since, for a given input dynamic range, more straight-line segments (thus more lin-limit transitions) are used to approximate the logarithmic function. The error from Figure 2-5 is the maximum error for a best least-mean-square curve fit (thus for a linear gain of 12 dB, the total error is 2 dB or ±1 dB for the least-mean-square logarithmic response).

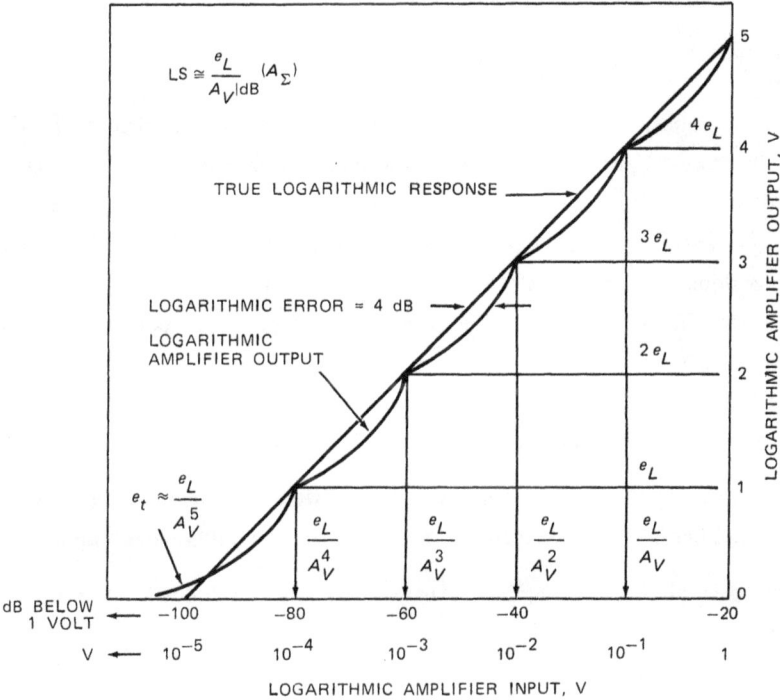

FIGURE 2-4. Characteristics of a Four-Stage Linear-Limiting Logarithmic Amplifier (Figure 2-3 and Equation 2-10). $A_V = 10$ (or 20 dB); $e_L = 1$ volt.

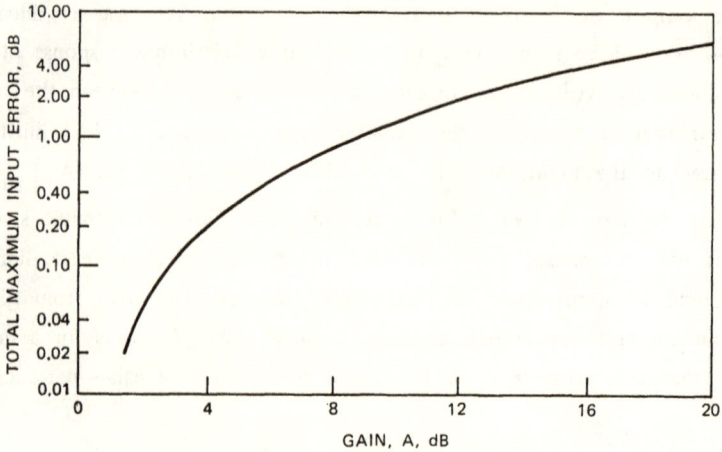

FIGURE 2-5. Total Logarithmic Error as a Function of Single-Stage Gain.

It was assumed that $A_V \gg 1$ to neglect the terms to the right of e_L in Equation 2-10; however, for large gains the logarithmic error is excessive (see Figure 2-4). In actual practice, the linear gains will be less than 15 dB (total error of $\pm 1\ 1/2$ dB), and the unwanted terms in Equation 2-10 cannot be neglected. The net effect is a decreasing slope for large inputs (since the output will decrease by e_L/A_V^n for each limited stage (Equation 2-10)). The decrease in slope is, of course, a function of A_V; the smaller the A_V, the greater the decrease in slope. This slope decrease is minimized by tapering the gain in the summing amplifier (more gain for the higher level stages) or by increasing e_L for the high-level stages.

This technique can be simplified if a differential amplifier is used as the limiting amplifier (Appendix A gives a discussion of the differential amplifier).

The basic differential amplifier (Figure 2-6) has the following transfer function (see Appendix A):

$$e_{out} = \frac{I_T\ R_C}{2} \tanh \left[\frac{e_{in}}{2V_T} \right] \qquad (2\text{-}11)$$

20

where

$$V_T = \frac{KT}{q}$$

$$\phi_1 = \phi_2 = V_{BE}$$

$$I_{C_1} = I_{C_2} \cong \frac{I_T}{2}$$

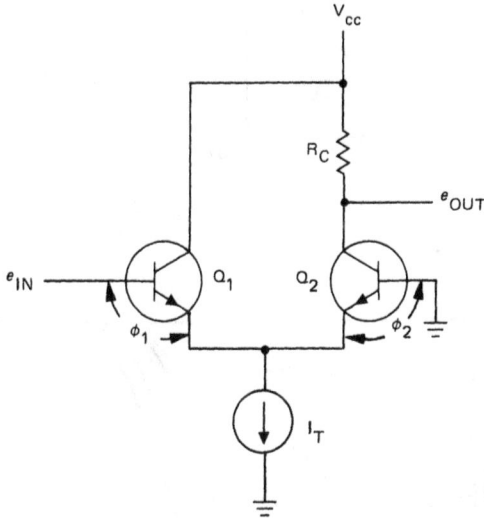

FIGURE 2-6. Basic Differential Amplifier.

Equation 2-11 assumes that Q_1 and Q_2 are identical and have equal base biases. This is a reasonable assumption for well-matched monolithic integrated transistor arrays [6-8].

It has been shown [see 3] that well-matched transistors follow Equation 2-11 nearly exactly. Figure 2-7 illustrates a normalized plot of Equation 2-11. Note that the amplifier has excellent dual-polarity limiting

$$e_L = \pm \frac{I_T R_C}{2} \qquad (2-12)$$

Figure 2-8 is a one-quadrant plot of Equation 2-11 for inputs in decibels at three temperatures.

21

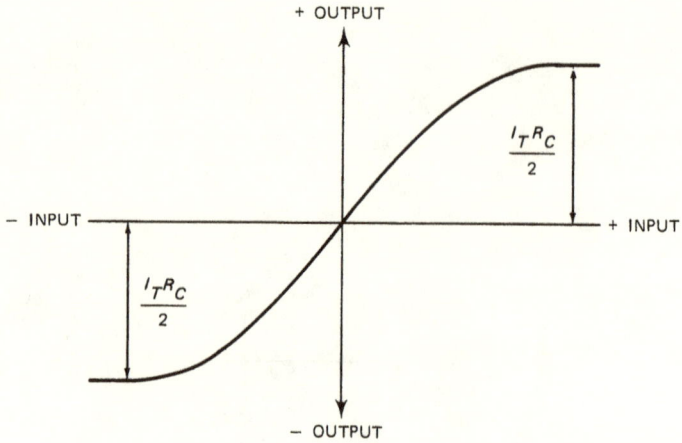

FIGURE 2-7. Plot of $\tanh \dfrac{e_{in}}{2V_T}$.

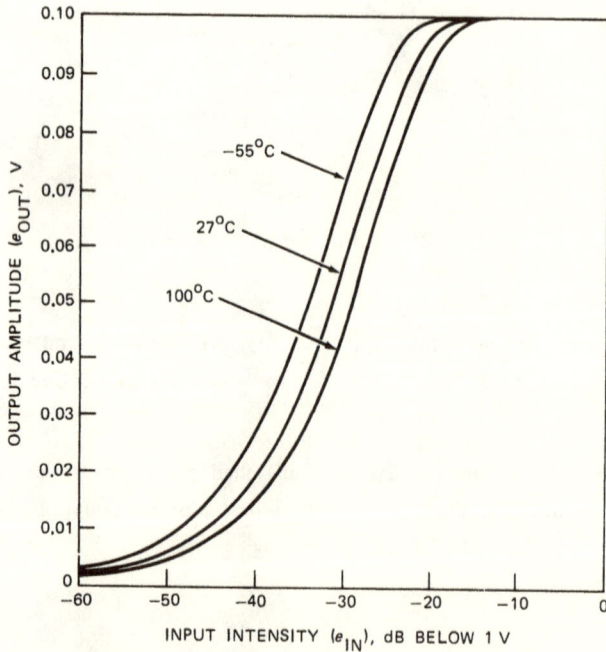

FIGURE 2-8. Plot of Equation 2-11 for $I_T R_C/2 = 0.1$ and $V_T = 0.027$.

Note several interesting features from Figures 2-7 and 2-8. The output has symmetrical limiting. The limit occurs for inputs of approximately −20 dB (or 0.1 V). The output is symmetrical and closely approximates a logarithmic function over a limited dynamic range.

It is possible to cascade stages in the same manner as for the lin–limit logarithmic amplifier and to obtain a reasonable logarithmic response. The basic concept is the same as for the lin–limit case; however, the equations are quite involved and yield little insight into the circuit's operation. A simpler method to analyze this configuration is to plot, via a digital computer, the output for various cascaded stages (varying the number of cascaded stages and the bias parameter B ($B = I_T R_C/2$)).

Figures 2-9 through 2-13 illustrate the effect of n (the number of cascaded stages) and B ($B = I_T R_C/2$) on the logarithmic response. Increasing the number of stages increases the slope and extends the dynamic range, but has no effect on the logarithmic error. Decreasing B decreases the dynamic range and the logarithmic error. The input is not summed in Figures 2-9 through 2-12. Figure 2-13 illustrates the logarithmic response for the input summed (n = 5 and various values for B). It is seen that summing the input extends the input dynamic range (by nearly 20 dB for B = 1). Figures 2-14 and 2-15 illustrate the temperature dependence on the logarithmic characteristic (increasing the temperature increases the slope and decreases the logarithmic error) while Table 2-1 summarizes the results obtained from Figures 2-9 through 2-13.

Comparing the results of Table 2-1 with Figure 2-5 shows that the differential amplifier nonlinear parallel summation logarithmic technique has, for B = 0.3, a dynamic range of 25 dB/stage and a logarithmic error of ±0.5 dB; while the dynamic range for the same error, using the lin–limit technique, is 8.5 dB/stage. Thus, for a given error, the nonlinear technique needs only one–half the stages needed for the lin–limit technique. This nonlinear differential amplifier logarithmic technique is a simple and most straightforward technique and may be easily implemented with integrated transistor arrays.

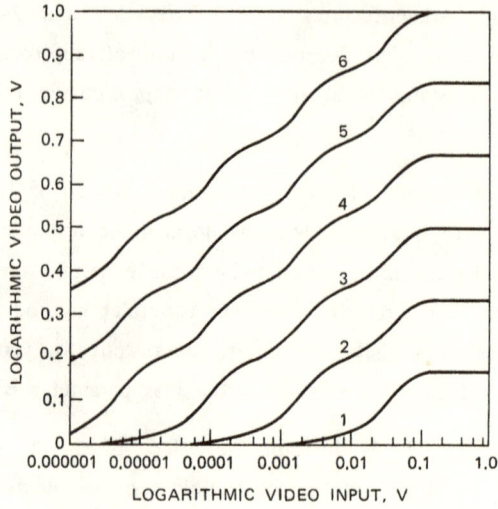

FIGURE 2-9. Output Versus Input for Cascaded Differential Amplifiers Where B = 1.0 and n = 1 to 6.

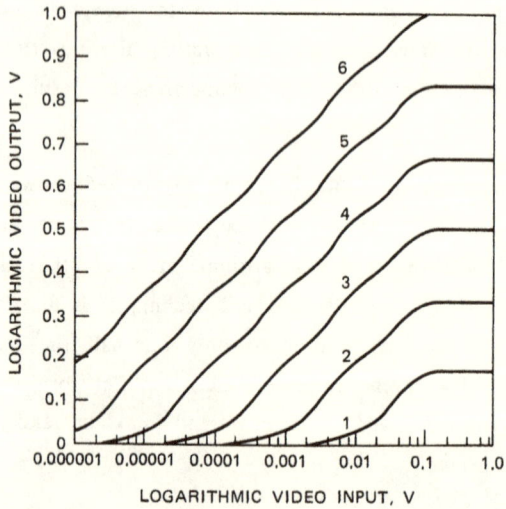

FIGURE 2-10. Output Versus Input for Cascaded Differential Amplifiers Where B = 0.5 and n = 1 to 6.

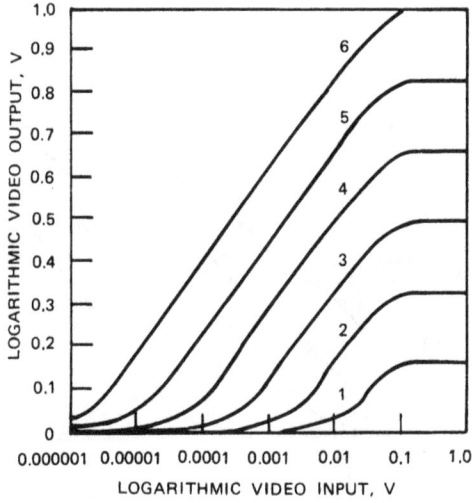

FIGURE 2-11. Output Versus Input for Cascaded Differentail Amplifiers Where $B = 0.3$ and $n = 1$ to 6.

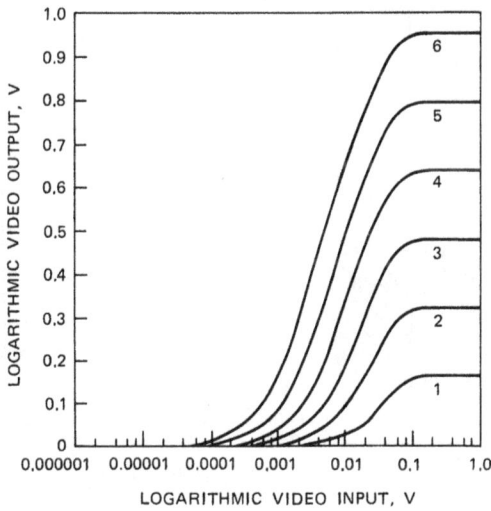

FIGURE 2-12. Output Versus Input for Cascaded Differential Amplifiers Where $B = 0.1$ and $n = 1$ to 6.

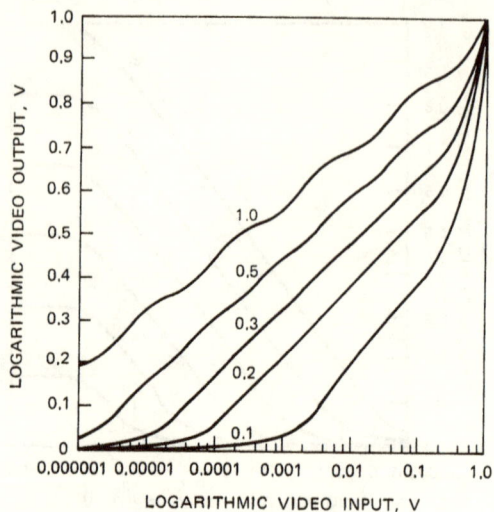

FIGURE 2-13. Output Versus Input for Cascaded Differential Amplifiers With Input Summed, With B = 0.1 to 1.0 and n = 5.

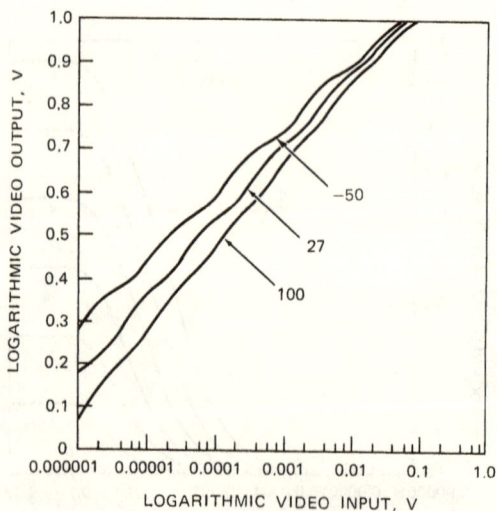

FIGURE 2-14. Output Versus Input for Cascaded Differential Amplifiers With Input Not Summed, Where B = 0.5, n = 6, and T = 50, 27, and 100°C.

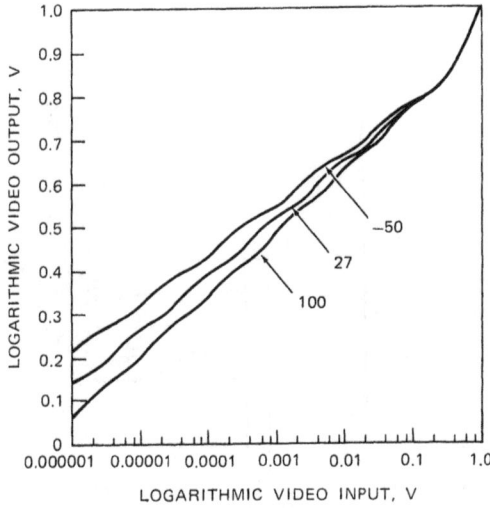

FIGURE 2-15. Output Versus Input for Cascaded Differential Amplifiers With Input Summed, Where $B = 0.5$, $n = 6$, and $T = -50$, 27, and $100°C$.

TABLE 2-1. Summary of Nonlinear Summation Logarithmic Amplifier.

Bias parameter	Dynamic range, dB/stage	Logarithmic error, dB
$B = 1$	25	±2
$B = 0.5$	20	±1
$B = 0.3$	15	±0.5
$B = 0.1$	7	±0.1

Figure 2-16 illustrates a four-stage differential nonlinear-summation logarithmic video amplifier. The constant B is varied by changing the –12-V supply on the constant-current transistors. Figure 2-17 illustrates the logarithmic characteristics for $B = 1$ and $B = 0.5$. The effect of summing the input is also shown. The results obtained from Figure 2-17 compare favorably with the predicted results of Figures 2-9 through 2-13.

27

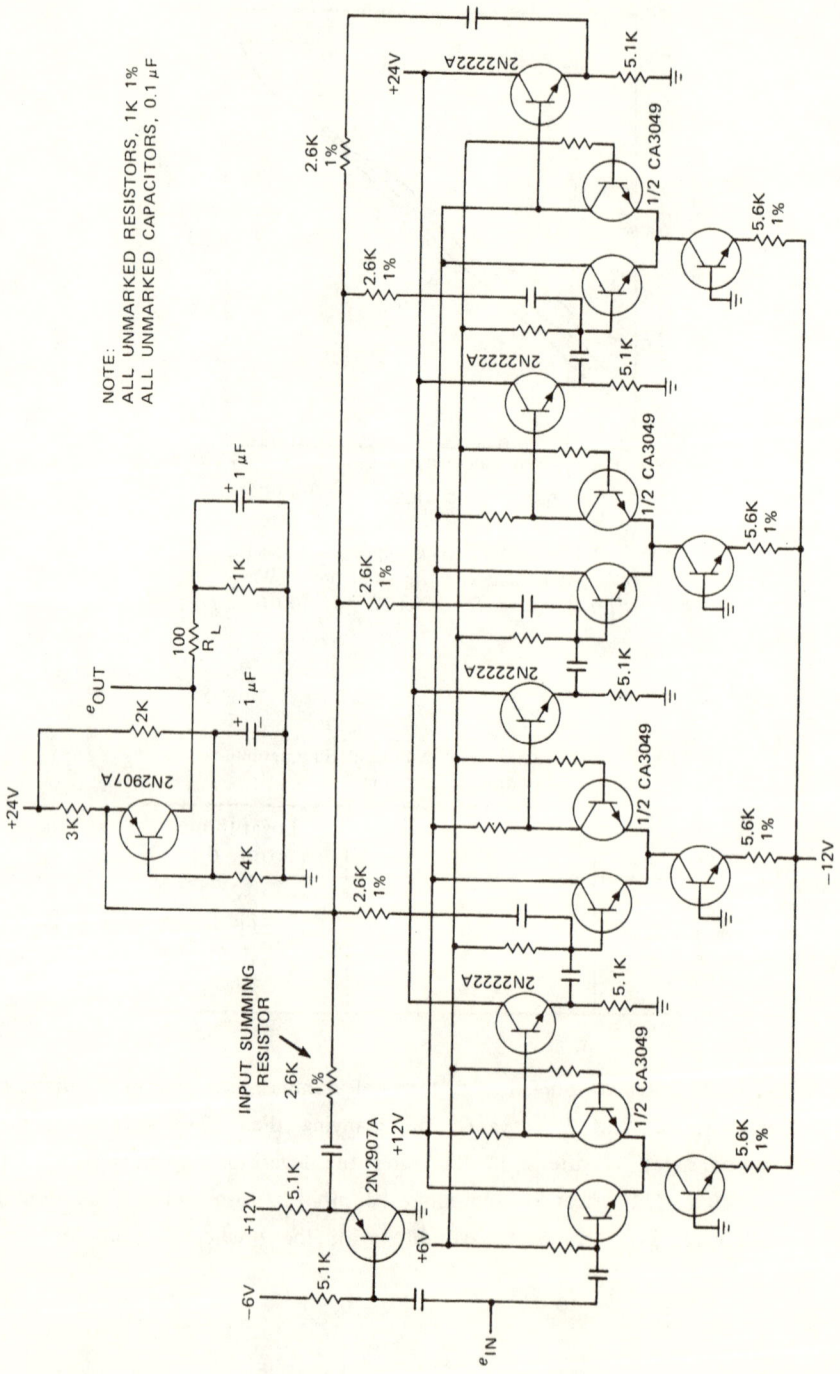

NOTE:
ALL UNMARKED RESISTORS, 1K 1%
ALL UNMARKED CAPACITORS, 0.1 μF

FIGURE 2-16. Four-Stage, Parallel, Nonlinear-Limiting, Logarithmic Video.

28

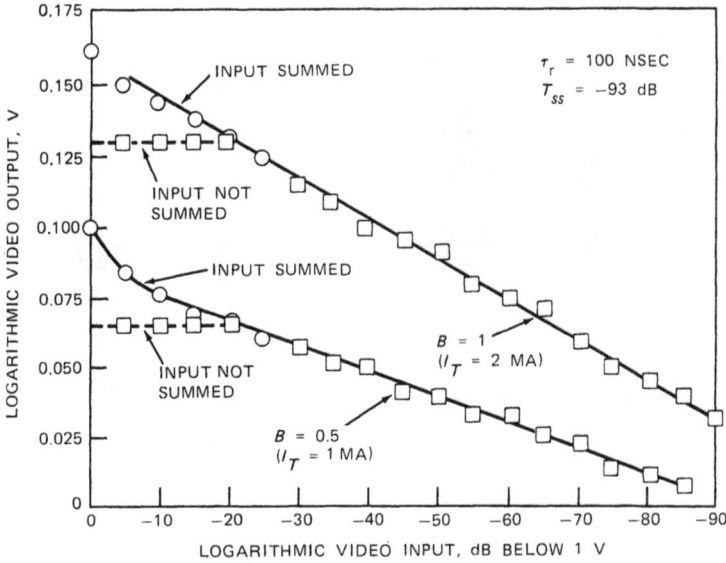

FIGURE 2-17. Characteristics of Four-Stage, Parallel, Nonlinear-Limiting, Logarithmic Video Amplifier.

Despite its simplicity, this technique has not gained general acceptance in the logarithmic community. This is due primarily to its change in LS with temperature (see Figures 2-14 and 2-15) and the difficulty in DC coupling stages (a topic covered in Chapter 4). A similar, series lin–limit, technique that is gaining attention is presented at the end of this chapter.

The two circuit techniques presented thus far are a parallel summation of series lin–limit and parallel nonlinear–limit stages. The next section presents a different parallel summation technique that is currently the most widely used logarithmic video technique.

Parallel Summation Logarithmic Video Amplifiers

Parallel nonlinear summation is illustrated in Figure 2-18. Logarithmic stages (L_n) are parallel to one another and driven by linear amplifiers (A_n).

29

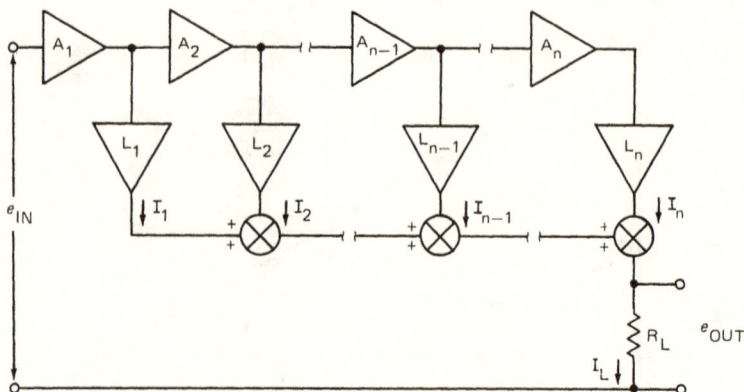

FIGURE 2-18. The General Form of a Multistage, Parallel
Summation, Logarithmic Video Amplifier.

The multistage amplifier (Figure 2-18) consists of n–log stages (L) with their output currents summed to give an approximation of a true logarithmic response over a wide dynamic range of inputs. Each stage of the generalized logarithmic amplifier consists of a linear amplifier (A) and a logarithmic amplifier (L) that has logarithmic input–output characteristics over a dynamic range specified by its transfer function.

The linear amplifiers are selected to phase in each log stage sequentially as a function of input intensity. The resultant composite output (Figure 2-19) approximates a true logarithmic response.

Figure 2-20 illustrates the output voltage for the basic differential amplifier logarithmic element (see Figure 2-7). The equation for the collector voltage is given as (see Appendix A)

$$e_0 = \frac{I_T R_L}{2} \tanh\left[\frac{e_{in}}{2V_T}\right] \qquad (2\text{-}13)$$

30

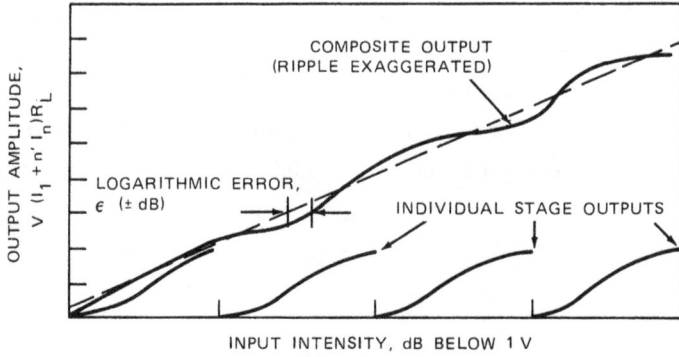

FIGURE 2-19. Composite Output of a Four-Stage, Parallel Summation, Logarithmic Amplifier.

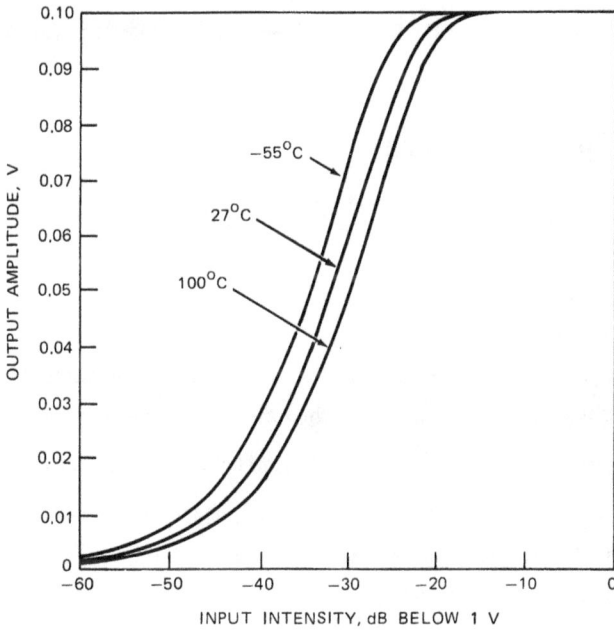

FIGURE 2-20. Output Versus Input for the Differential Amplifier Logarithmic Stage, Where $\dfrac{I_T R_C}{2} = 0.1$.

31

The output voltage is a close approximation to a logarithmic function for input voltages from -40 dB (10 mV) to -20 dB (100 mV) with the center of logarithmic action at -28 dB [see 3]. This logarithmic center, $\mathcal{C}$ is an important design parameter as will be shown. If one wishes to use, say, 20 dB input dynamic range per differential amplifier logarithmic stage, the minimum input for logarithmic action is given as

$$e_{in}\bigg|_t^{dB} = \mathcal{C}(dB) - \frac{dB/stage}{2} \tag{2-14}$$

or, for 20 dB/stage

$$e_{in}\bigg|_t^{dB} = -28 - 20/2 = -38 \ dB \tag{2-15}$$

Thus, each 20-dB logarithmic stage will start its logarithmic action at -38 dB, with each logarithmic stage offset by a 20-dB gain. The LS (in V/dB) is given as [see 3]

$$LS = \frac{I_T R_L}{2} \left(\frac{1}{dB/stage}\right) \ volts/dB \tag{2-16}$$

or in terms of current

$$LS(current) = \frac{I_T}{2} \left(\frac{1}{dB/stage}\right) \frac{amp}{dB} \tag{2-17}$$

The design process is basically quite simple and is illustrated in Table 2-2. Figures 2-21 through 2-24 illustrate the theoretical response for several configuations $\left(e_{in}\big|_t^{dB} = -90 \ dB \ and \ e_{in}\big|_{dB}^{max} = +10 \ dB\right)$. The results predicted using Table 2-2 are excellent; however, as can be seen, there is a shift of the logarithmic curve (LS remaining constant) with temperature. This is usually of minor concern; however, the cause and cure are discussed in Appendix B.

TABLE 2-2. Parallel Summation Logarithmic Video Amplifier Design Procedure.

A. Determine where logarithmic action is to start, $e_{in}\Big|_{t}^{dB}$ and the maximum input for logarithmic action, $e_{in}\Big|_{dB}^{max}$ (or input dynamic range).

B. Determine the needed logarithmic conformity (see Appendix B)

Logarithmic error, ±dB	dB/stage
0.8	20
0.4	15
0.2	10

C. Find the number of logarithmic stages, n

$$n = \frac{\left| e_{in}\Big|_{t}^{dB} + e_{in}\Big|_{dB}^{max} \right|}{dB/stage}$$

D. Give the LS as

$$LS = \frac{I_T R_L}{2}\left(\frac{1}{dB/stage}\right)$$

and $\frac{I_T R_L}{2}$ is chosen for the wanted LS.

E. Determine the voltage gain needed preceding the lowest level logarithmic stage (see Equation 2-13).

$$A_V\Big|_{dB} = -28 - \frac{dB/stage}{2} - e_{in}\Big|_{t}^{dB}$$

or

$$e_{in}\Big|_{t}^{dB} = -A_V\Big|_{dB} - \frac{dB/stage}{2} -28$$

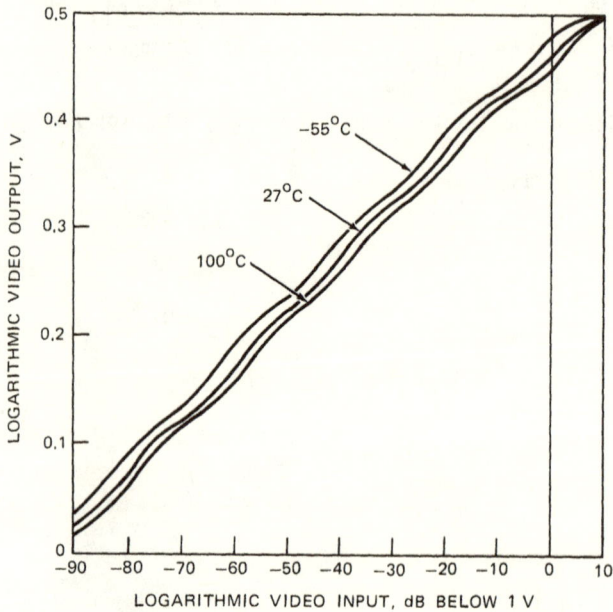

OUTPUT VERSUS INPUT FOR THREE TEMPERATURES

GAIN NEEDED PRECEDING LOGARITHMIC STAGE

STAGE	GAIN, dB
1	−28
2	−8
3	12
4	32
5	52

LS (Table 2-2) = 5 mV/dB $\qquad$ LS (MEASURED) = 4.9 mV/dB

$e_{IN}\left.\right|_t^{dB}$ (Table 2-2) = −90 dB $\qquad$ $e_{IN}\left.\right|_t^{dB}$ (MEASURED) = −90 dB

FIGURE 2-21. 20-dB/Stage, Five-Stage Parallel
Summation Logarithmic Video $\left(\dfrac{I_T R_L}{2} = 0.1\right)$.

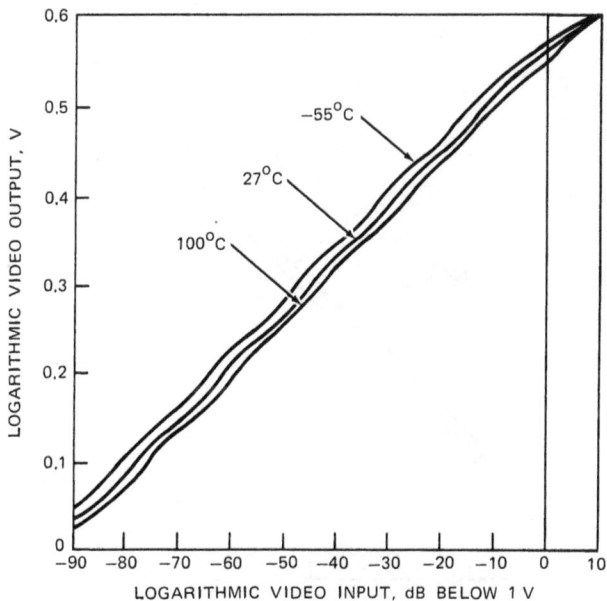

OUTPUT VERSUS INPUT FOR THREE TEMPERATURES

GAIN NEEDED PRECEDING LOGARITHMIC STAGE

STAGE	GAIN, dB
1	−29.8
2	−13.1
3	3.6
4	20.3
5	37.0
6	53.7

LS (Table 2-2) = 6 mV/dB LS (MEASURED) = 5.9 mV/dB

$e_{IN}\big|_t^{dB}$ (Table 2-2) = −90 dB $e_{IN}\big|_t^{dB}$ (MEASURED) = −90 dB

FIGURE 2-22. 16.7-dB/Stage, Six-Stage Parallel Summation Logarithmic Video $\left(\dfrac{I_T R_L}{2} = 0.1\right)$.

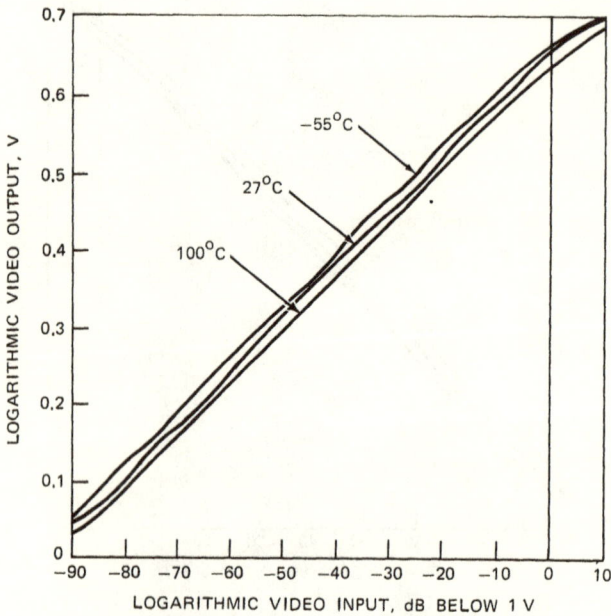

OUTPUT VERSUS INPUT FOR THREE TEMPERATURES

GAIN NEEDED PRECEDING LOGARITHMIC STAGE

STAGE	GAIN, dB
1	−30.9
2	−16.6
3	−2.3
4	12.0
5	26.3
6	40.6
7	54.9

LS (Table 2-2) = 7 mV/dB LS (MEASURED) = 6.9 mV/dB

$e_{IN}\Big|_t^{dB}$ (Table 2-2) = −90 dB $e_{IN}\Big|_t^{dB}$ (MEASURED) = −90 dB

FIGURE 2-23. 14.3-dB/Stage, Seven-Stage Parallel Summation Logarithmic Video $\left(\dfrac{I_T R_L}{2} = 0.1\right)$.

36

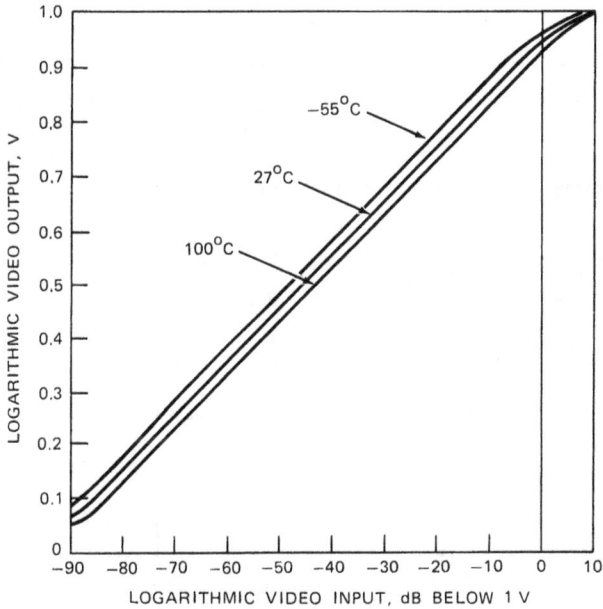

OUTPUT VERSUS INPUT FOR THREE TEMPERATURES

GAIN NEEDED PRECEDING LOGARITHMIC STAGE

STAGE	GAIN, dB
1	−33
2	−23
3	−13
4	−3
5	7
6	17
7	27
8	37
9	47
10	57

LS (TABLE 2-2) = 10 mV/dB LS (MEASURED) = 9.6 mV/dB

$e_{IN}\Big|_t^{dB}$ (TABLE 2-2) = −90 dB $e_{IN}\Big|_t^{dB}$ (MEASURED) = −90 dB

FIGURE 2-24. 10-dB/Stage, Ten-Stage Parallel Summation Logarithmic Video $\left(\dfrac{I_T R_L}{2} = 0.1\right)$.

37

Note that for low dB/stage values (see Figures 2-23 and 2-24), the logarithmic response deviates from a straight line approximately n dB per stage (where n is the number of log stages) before the specified maximum input. To minimize the effect, the input is summed as in Figure 2-3; however, as will be shown, adding another differential stage to act as a high-level summer is a simple and practical solution.

The following discussion of a practical logarithmic video amplifier is related to the theory presented. This configuration (Figures 2-25 and 2-26) is a basic building block for much of what follows. The intent is to design as much of the logarithmic amplifier with integration (master chip monolithic integrated circuit format) in mind [see 6-8]. Various circuit design trade-offs are noted as we proceed.

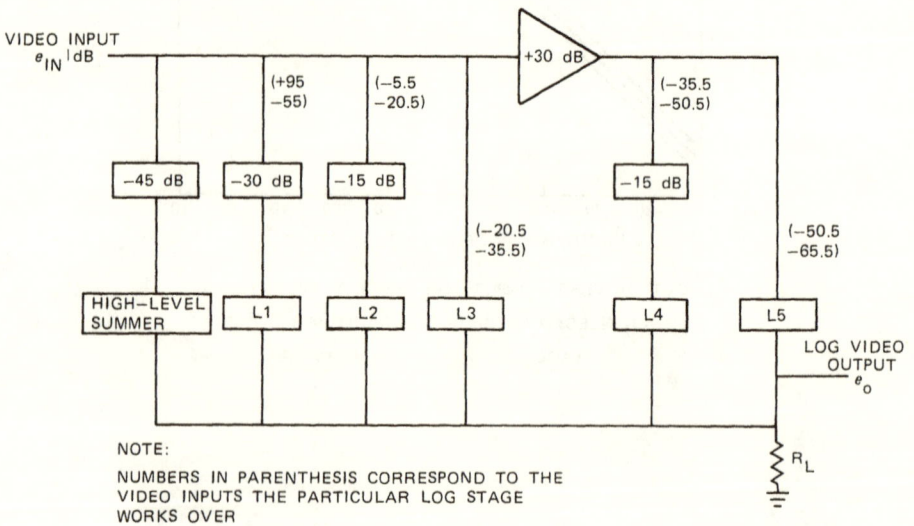

FIGURE 2-25. Basic Five-Stage, 15-dB/Stage, Parallel Summation, Logarithmic Video With High Level Summer.

Figure 2-25 is a block diagram of the basic 15-dB/stage design (with high level summer); a circuit diagram is given in Figure 2-26. Each logarithmic stage has an equal bias resistance from each base to the ground. This ensures equal V_{BE} voltages, as a 5-mV difference in base voltage gives a 1.44-dB error* (see Appendix B).

*This is most important for DC-coupled logarithmic stages where one must be concerned with the voltage offsets and drift with temperature for the linear amplifiers driving the logarithmic stages.

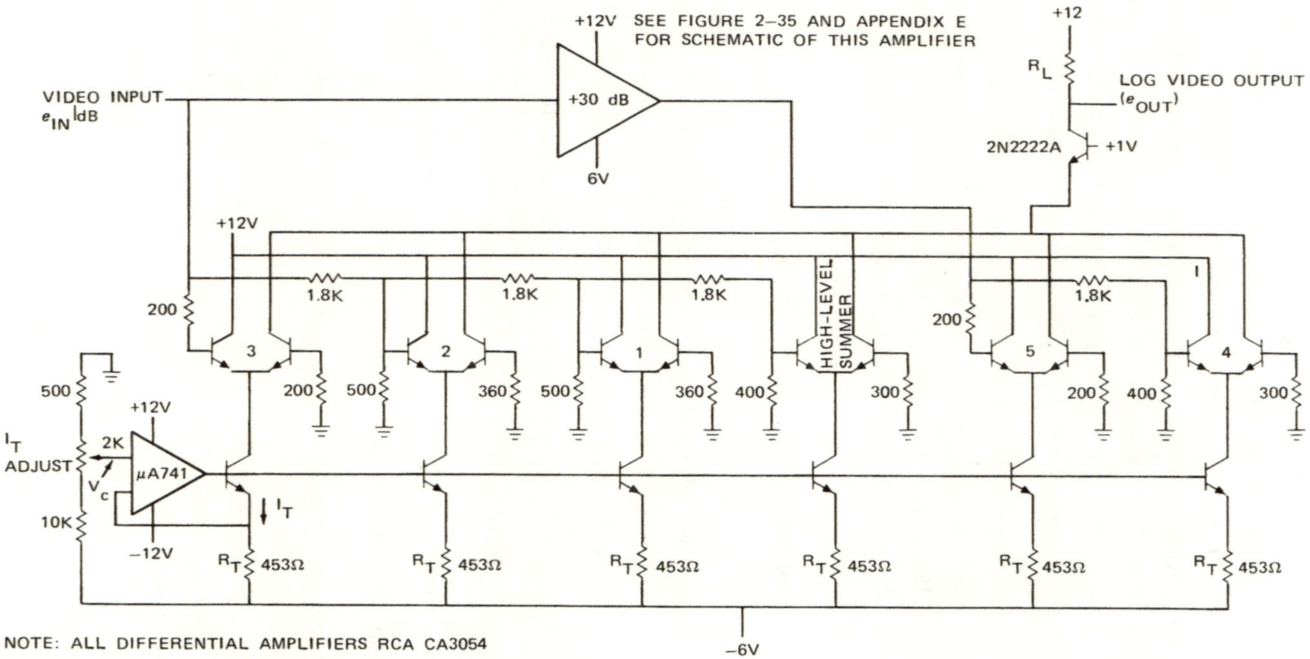

NOTE: ALL DIFFERENTIAL AMPLIFIERS RCA CA3054

$\Delta V_{BE} < 0.5$ mV $\quad V_{EBO} = 7$ VOLTS

FIGURE 2-26. 15-dB/Stage, Five-Stage, Parallel Summation, Logarithmic Video Amplifier With High Level Summer.

The current, I_T, is set via the I_T adjust as shown. This is not necessarily an optimum current generator and is merely used for simplicity. Appendix C discusses various constant current sources suitable for use if the logarithmic stages are contained in a well-matched monolithic integrated circuit. To minimize the effective capacitance associated with summing the six collectors together, the cascode configuration as shown is used (the savings in pulse response will be illustrated momentarily). Appendix D presents several output stages that have proven useful.

Figure 2-27 shows the measured results for various values of I_T and R_L. The amplifier has an excellent logarithmic response over a 65-dB region with no high-level summer and 75-dB with the high-level summer, with a logarithmic error less than ±1/2 dB (which is to be expected (see Appendix B)). Table 2-3 compares the predicted and measured results, which are quite close. Extending the log transfer function until it meets the e_{in} axis $\left(e_{in}\Big|_{dB}^{e_o=0}\right)$, $-K_2\Big|_{dB}$ (see Figure 1-2) is obtained. It is independent of I_T or R_L (the LS) and is, in fact, only dependent upon the input necessary to exceed the lin-log transition of the lowest level logarithmic stage (Equation 2-14) and to a lesser degree on temperature (see Figures 2-21 to 2-24 and Appendix B). This is most important as the LS has no effect on where the amplifier enters its logarithmic region. Decreasing the input for $e_{in}\Big|_t^{dB}$ may be accomplished (with no change in LS) by placing an amplifier at the input. A 20-dB gain decreases the lin-log transition $e_{in}\Big|_t^{dB}$ by 20 dB (the logarithmic response of Figure 2-27 merely shifts 20 dB to the left).

The pulse response (for inputs from -20 to -60 dB) for an R_L = 750, with and without the cascode amplifier, is shown in Figure 2-28. The effect of the summed parallel capacitance of the logarithmic stage is easily seen. If a large LS is needed for a given application, the cascode (or an equivalent technique) may well be required (see Appendix D).

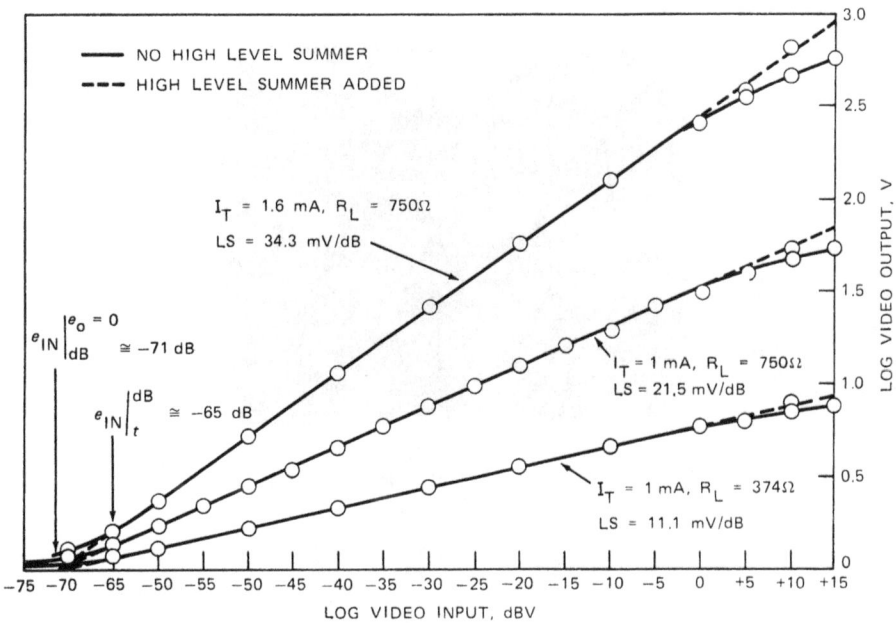

FIGURE 2-27. Five-Stage Parallel Summation Logarithmic Video Amplifier Results (15-dB/Stage).

TABLE 2-3. Measured and Predicted Result Summary Five-Stage Log Video.

Parameter		Measured	Predicted (see Table 2-2)
Start of logarithmic action $\left(e_{in}\Big|_t^{dB}\right)$		-65 dB	-65.5 dB
Log slope (LS)	$I_T = 1.6$ mA* $R_L = 750$	34.3 mV/dB	40 mV/dB
	$I_T = 1$ mA* $R_L = 750$	21.5 mV/dB	25 mV/dB
	$I_T = 1$ mA* $R_L = 374$	11.1 mV/dB	12.5 mV/dB
Instantaneous input dynamic range		75 dB (with summer)	75 dB

*mA = milliamperes

41

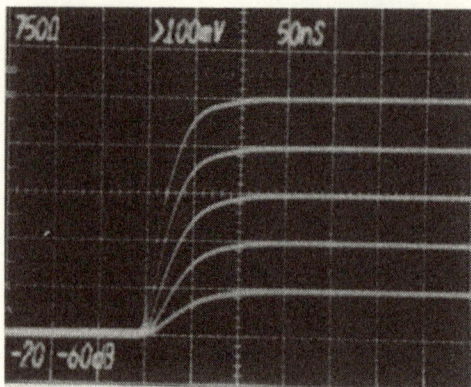

(a) I_T = 1 mA, R_L = 750Ω, No Cascode Output
-20 to -60 dB in 10-dB Steps.

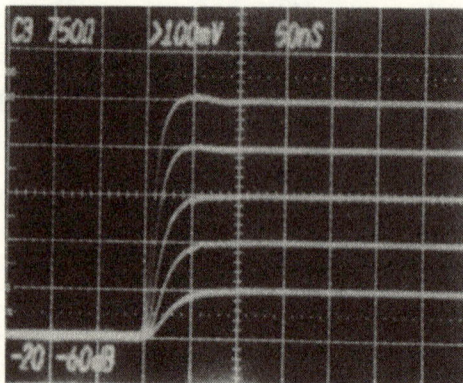

(b) I_T = 1 mA, R_L = 750Ω, Cascode Output
-20, to -60 dB in 10-dB Steps.

FIGURE 2-28. Pulse Response for the Five-Stage,
15 dB/Stage Logarithmic Video.

Figure 2-29 illustrates the pulse response for the five-stage logarithmic amplifier for inputs from 0 to -65 dB. Note that the recovery time is excellent until the output drops to -35 dB referred to the input. The recovery time then takes several microseconds to return to the baseline. The cause of the problem is that either the input does not turn off as a step function or there is some decay associated with the linear video amplifier. This problem is illustrated in Figure 2-30. The top trace is the output of the linear video amplifier (off screen for inputs greater than -40 dB) showing the low-level recovery problem. The lowest level logarithmic stage (stage 5, Figure 2-26) logs for inputs greater than -35.5 dB (16.8 mV); thus, any low-level tail on the linear video will be logged until it is less than 16.8 mV.*

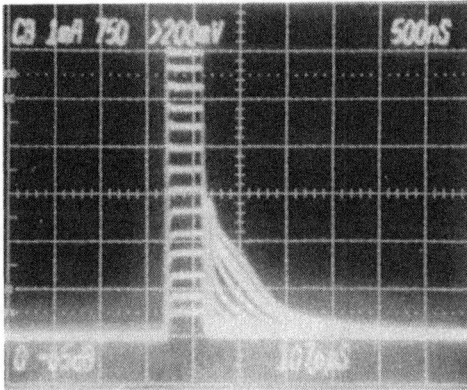

FIGURE 2-29. Pulse Response for the Five-Stage
Logarithmic Video, Input 0 to -65 dB in
5-dB Steps.

The above discussion points out a major concern in designing and testing logarithmic amplifiers. If we wish to test a logarithmic amplifier that has a dynamic range from -65 dB (562 μV) to +15 dB (5.62 V), the signal source must recover to at least -65 dB in a short time, or this will limit the recovery time. Stated a bit differently: If a test signal has "garbage" riding on the baseline greater than 562 μV (-65 dB), the logarithmic video will treat them like any other signal. This can drive one quite mad.

*This recovery problem is generally due to the recovery time at the linear video amplifier.

43

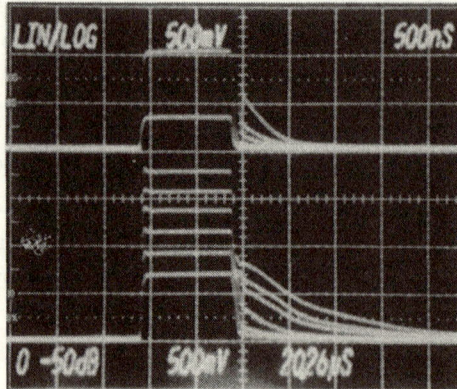

Top-Output of Linear Video Amplifier
(High Level Inputs of the Screen)
Bottom-Log Video Output. Inputs
from 0 to -50 dB in 10-dB Steps.

FIGURE 2-30. Pulse Recovery for the
Five-Stage Video.

When charged, a capacitor has an exponential decay, and the logarithm of an exponential is a straight line. Figure 2-31 illustrates this condition. If the falling slope is subtracted from a linear ramp [see 1], a straight line should be obtained. Any deviation from a straight line is an error with respect to ideal. Figure 2-32 illustrates the results of such a test. The error is within ±1/2 dB.

Detector-Logarithmic Video Amplifier

Figure 2-33 presents the detector configuration used [9]. (Appendix F presents a basic review of detector theory.) The tunnel diode* (Figure 2-34) is well-suited for detector-logarithmic video applications if the upper dynamic range limit is +5 dBm or less. Figure 2-34 is a normalized plot of the detector/amplifier output as a function of input power. Note the output is square law (increases 20 dBV for each 10 dBm increase in input power) for power inputs of -10 dBm or less. Shortly we will see that the nonlinear behavior above -10 dBm can easily be corrected by increasing the current through the high-level summer stage.

*Tunnel diode detectors are now available using planar processing, thus greatly improving the mechanical characteristics of the older mesa structure. Also temperatures in excess of +100°C can be handled [10].

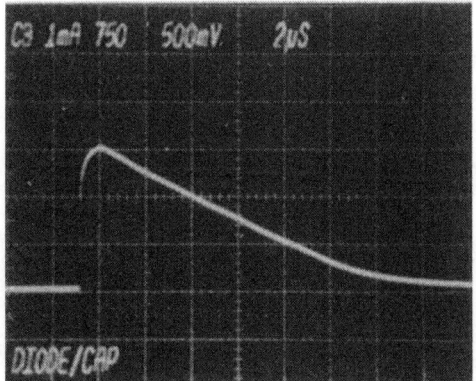

LOG SLOPE = 21:5 mV/dB, PEAK VALUE = 0 dBV INPUT

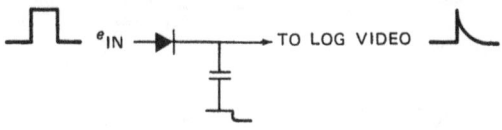

FIGURE 2-31. Logarithmic Response of an
Exponential Decay.

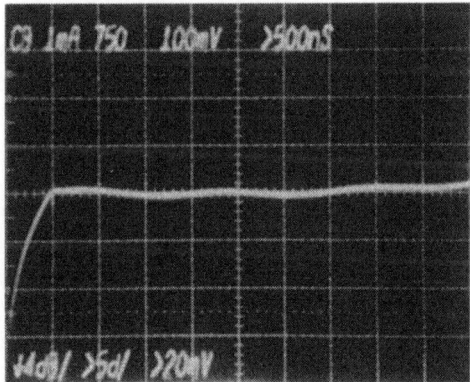

VERTICAL = 4 dB/DIV
HORIZONTAL = 5 dB/DIV (0 TO −40 dB)

FIGURE 2-32. Logarithmic Error.

45

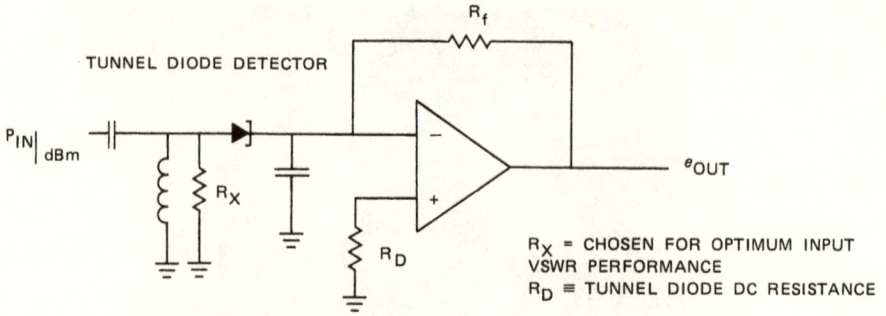

FIGURE 2-33. Tunnel Diode Detector/Video Amplifier
(see Appendix F for Details.)

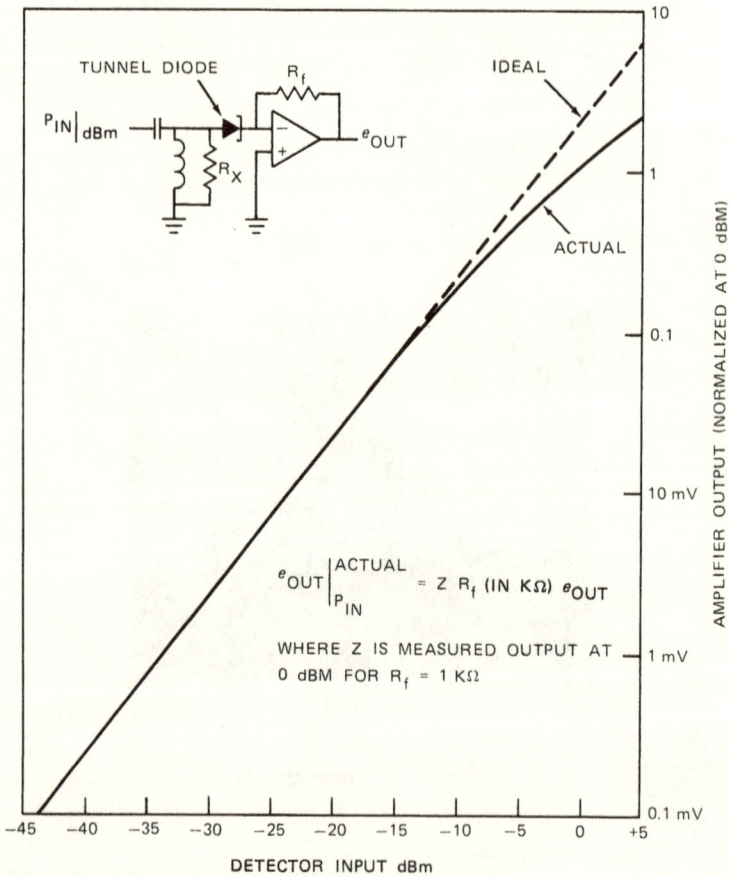

FIGURE 2-34. Normalized Detector Output
Versus Input Power.

46

The absolute lower limit for logarithmic action is the tangential signal sensitivity (T_{ss}) of the detector–linear video amplifier combination [11]. Appendix G presents a basic discussion on T_{ss} and its measurement.

The marriage of the detector–linear video amplifier (Figure 2-33) to the logarithmic video amplifier (Figure 2-26) will be covered momentarily. However, a few words on coupling. As mentioned previously (and covered in detail in Appendix B), any DC offset on the output of either linear amplifier distorts the logarithmic transfer function. The quest for true DC coupling is covered in Chapter 4. For now, it is assumed that there is no DC offset at the linear video outputs.

The T_{ss} for the tunnel diode/linear video amplifier is –43 dBm* (linear video bandwidth $\cong$ 7 MHz). As mentioned in Chapter 1, it is usually wise to have several decibels of linear amplification before the logarithmic response begins. Thus, the input for logarithmic action, $P_{in}\big|_t^{dBm}$ (Figure 1-5) is chosen as –39 dBm.

The normalizing factor, Z, for the tunnel diode chosen is 3.5, or the video output is given as (see Figure 2-34)

$$e_{out}\bigg|_{P_{in}}^{actual} = 3.5\ R_f\ (in\ k\Omega)\ e_{out}\big|_{normalized} \qquad (2\text{-}18)$$

The linear video amplifier used with the detector supplies 6 V easily. As this is fairly close to the base–emitter breakdown voltage of the logging stages (some 7 V), a value of 5.5 V is used for the maximum output at +5 dBm input. Thus, from Equation 2-18

$$R_f(in\ k\Omega) = \frac{e_{out}\big|_{+5\ dBm}}{Z\ e_{out}\big|_{normalized}} \qquad (2\text{-}19)$$

---

*In Figure 2-33, R_x is chosen to optimize the input voltage standing wave ratio. This also results in a T_{ss} loss of approximately 3 dB (T_{ss} without R_x = –47 dBm).

or

$$R_f = \frac{5.5}{3.5\ (2.1)} = 748\Omega \qquad\qquad (2\text{-}20)$$

A value of 750Ω will be used. The detector output for an input power of -39 dBm is (from Figure 2-34 and Equation 2-18)

$$e_{out} = 3.5\ (750)(0.3\ mV) = 0.79\ mV \qquad\qquad (2\text{-}21)$$

or in decibels

$$e_{out} = 20\ \log\ (0.79\ mV) \cong -62\ dB \qquad\qquad (2\text{-}22)$$

and this is quite close to the -65.5 dB input needed for the lowest level logarithmic stage (Figure 2-25).* Figure 2-35 illustrates the logarithmic configuration; Figure 2-36 shows the logarithmic transfer function. The response, for the current through the high-level summer equal to the logarithmic stage bias currents, deviates from the wanted logarithmic response at large input powers. Referring to the normalized detector response of Figure 2-34, for an input power of $+5$ dBm, the wanted output is 3 V, not 1 V. Thus, if a gain of 3 is provided in the high-level summer, the overall wanted response should be fairly close. The low-level amplification (inputs below -40 dB) for a differential is given as (see Appendix A)

$$A_V = \frac{I_T\ R_C}{4V_T} \qquad\qquad (2\text{-}23)$$

Thus, increasing I_T for the high level summer by a factor of 3 increases its low-level gain by the same amount. Increasing I_T for the high-level summer is easily accomplished by decreasing its current determining resistor. Figure 2-36 illustrates the logarithmic transfer function for $R_E = 150\Omega$.** The scale factor may be changed by changing R_L or I_T (keeping the logarithmic bias resistors at 453Ω and the high-level summer at 150Ω) with no change in logarithmic accuracy.

---

*This is obviously no accident, and I must confess this logarithmic configuration (Figures 2-25 and 2-26) is optimized for the normalized detector curve of Figure 2-34.

**In integrated circuit form, the bias current for the high-level summer is actually three paralleled transistor current sources with each $R_E = 453\Omega$ This guarantees a 3:1 current ratio over temperature.

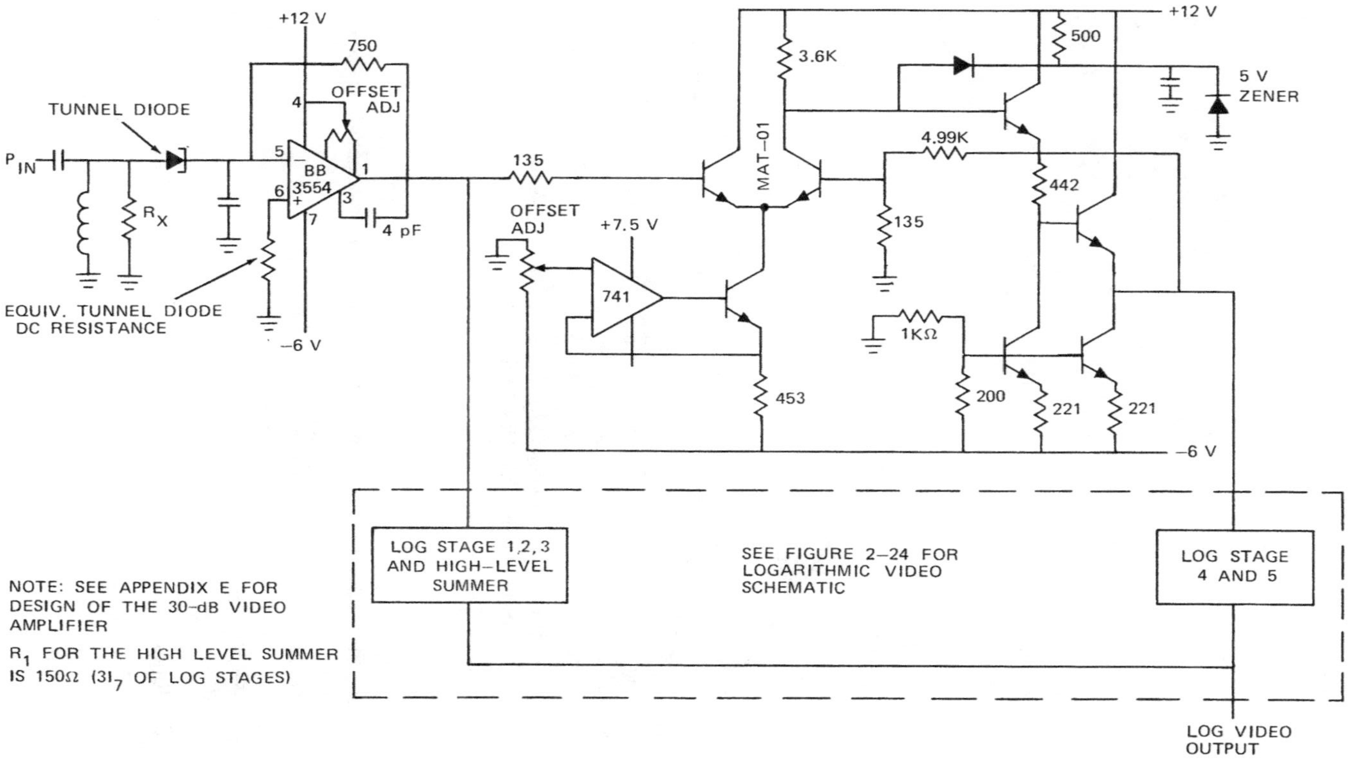

FIGURE 2-35. Detector-Logarithmic Video Amplifier.

49

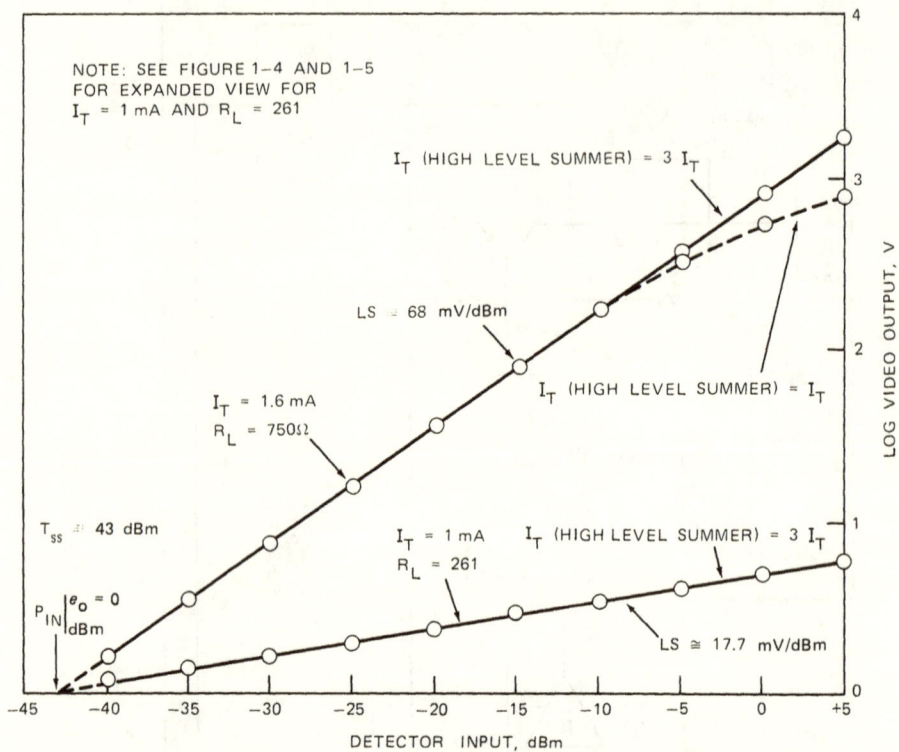

FIGURE 2-36. Detector–Logarithmic Amplifier Results
Five-Stage Logarithmic Plus High-Level Summer.

The LS for the detector–logarithmic video is given as (assuming square law detection)

$$LS = \frac{I_T\ R_L}{dB/stage}\ V/dBm \tag{2-24}$$

The measured slopes for Figure 2-36 are 68 and 17.7 mV/dBm with predicted values of 80 and 17.4 mV/dBm, respectively.

Figure 2-37 shows the pulse response for inputs from +5 to -40 dBm (the negative spike at the leading edge of Figure 2-37a is test equipment-induced). The recovery time is shown in Figure 2-37b, and it is seen that the output drops fairly quickly to -25 dBm. The increase in recovery time from this point is primarily due to the recovery time of the 30-dB linear video amplifier (see Figures 2-29 and 2-30).

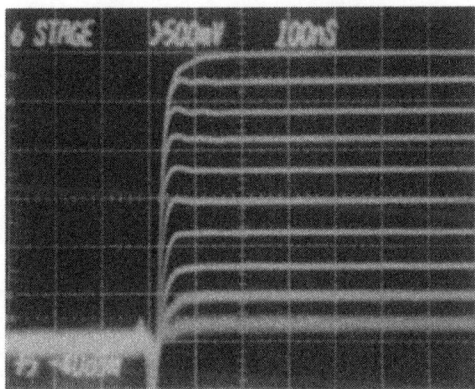

(a) Effect of Input Intensity on Leading Edge
(+5 to -40 dBm in 5-dB Steps).

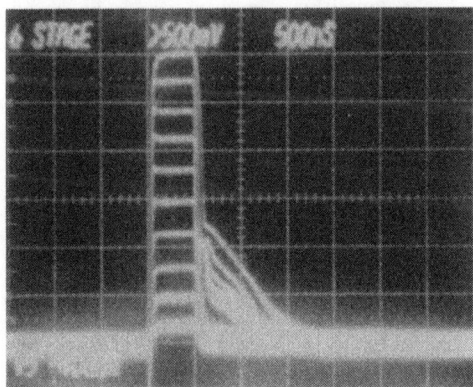

(b) Effect of Input on Recovery
(+5 to -40 dBm in 5-dB Steps).

FIGURE 2-37. Pulse Response for Detector-
Logarithmic Video Amplifier.

51

The technique just presented is the one predominantly used in radar and EW detector-logarithmic video amplifiers. If narrow pulses (less than 30 nsec or so) need to be logged, the series lin-limit technique offers distinct advantages [see 4,12] and pulse widths as low as 9 nsec can be handled [13].

Series Linear-Limit Logarithmic Video Amplifier

Figure 2-3 illustrates the parallel lin-limit logarithmic technique with its characteristics illustrated in Figure 2-4. The same result can be accomplished by cascading in series (thus avoiding the delays associated with the parallel stages sequentially limiting). The series lin-limit circuit is shown in Figure 2-38. Barber and Brown [14] show the output, e_{out}, may be given as

$$e_{out} = e_L \left[N + \frac{1}{A} + \frac{Log\left(\frac{A e_{in}}{e_L}\right)}{Log\ (A + 1)} \right] \tag{2-25}$$

This may be rewritten as

$$e_{out} = e_L N + \frac{e_L}{A} - \frac{e_L}{Log\ (A + 1)} \left[Log\left(\frac{A}{e_L}\right) \right] + \frac{e_L}{Log\ (A + 1)} Log\ e_{in} \tag{2-26}$$

Equation 1-10 shows that a logarithmic response may be given as

$$e_{out} = K_1\ Log\ e_{in} + K_1\ Log\ K_2$$

Thus

$$K_1\ Log\ K_2 = e_L\ N + \frac{e_L}{A} - \frac{e_L}{Log\ (A + 1)} \left[Log\left(\frac{A}{e_L}\right) \right] \tag{2-27}$$

$$K_1 = \frac{e_L}{Log\ (A + 1)} \tag{2-28}$$

Letting

$$K_1\ Log\ K_2 = K_3 \tag{2-29}$$

$$e_{out} = \frac{e_L}{Log\ (A + 1)} Log\ e_{in} + K_3 \tag{2-30}$$

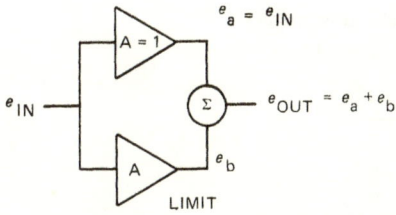

(a) Single lin-limit stage.

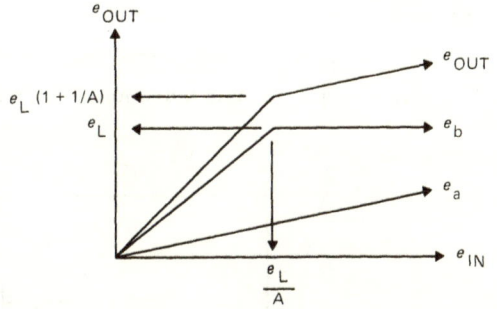

(b) Single lin-limit characteristic.

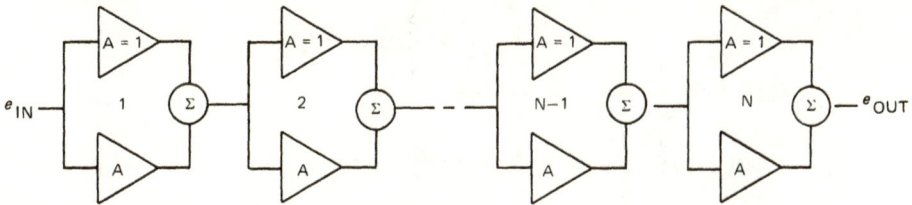

(c) Cascaded series lin-limit logarithmic video amplifier.

FIGURE 2-38. Basic Linear-Limiting Logarithmic Video Characteristics.

The input, e_{in}, is given in terms of decibels as

$$e_{in} = 10^{\frac{e_{in|dB}}{20}}$$ (2-31)

Substituting Equation 2-31 into 2-30

$$e_{out} = \frac{e_L}{20 \text{ Log } (A + 1)} \, e_{in|dB} + K_3$$ (2-32)

where

$$\frac{e_L}{20 \text{ Log } (A + 1)} = LS \text{ in volts/dB}$$ (2-33)

Figure 2-39 illustrates a three-stage series lin-limit logarithmic video to illustrate the pulse responses available with this technique. The differential amplifier is used as

53

the limiting stage. Since the differential amplifier gain is nonlinear (Equation 2-11), Equation 2-26 is not completely valid; however, the error is mainly associated with K_3, with the LS given in Equation 2-33 relatively unchanged.

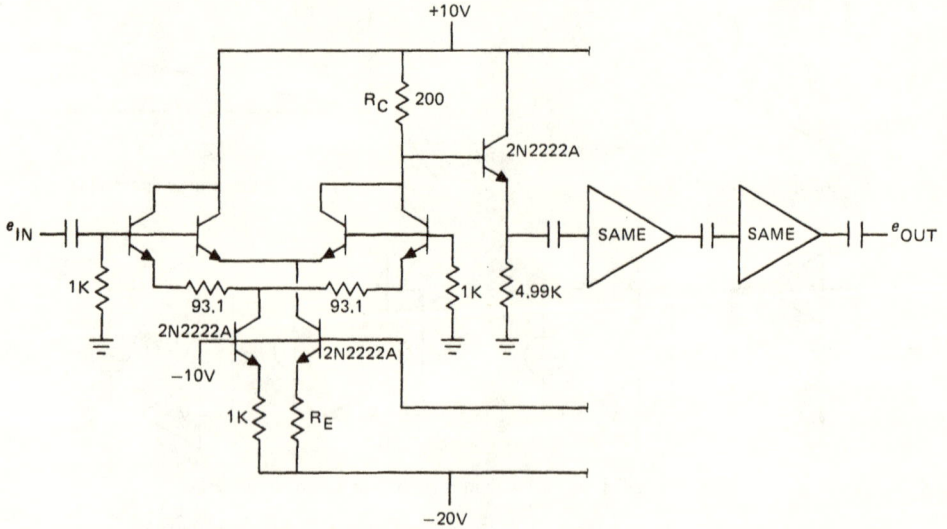

NOTE:

R_E SETS I_T FOR THE LIMITING STAGE

$$e_L = \frac{I_T R_C}{2}$$

ALL UNMARKED TRANSISTORS MD–918

FIGURE 2-39. Three-Stage Series Linear-Limiting Logarithmic Video Amplifier.

The value for A in Equation 2-33 is the differential amplifier low–level signal gain.

$$A = \frac{I_T R_C}{4V_T} \qquad (2\text{-}34)$$

Figure 2-40 illustrates the transfer function for several values at I_T, and Figure 2-41 shows the leading edge pulse response for $I_T = 4$ mA. The output reaches the peak value in less than 20 nsec. This configuration is inherently bipolar due to the differential amplifier and is a basic building block for the true logarithmic IF discussed in the next chapter.

54

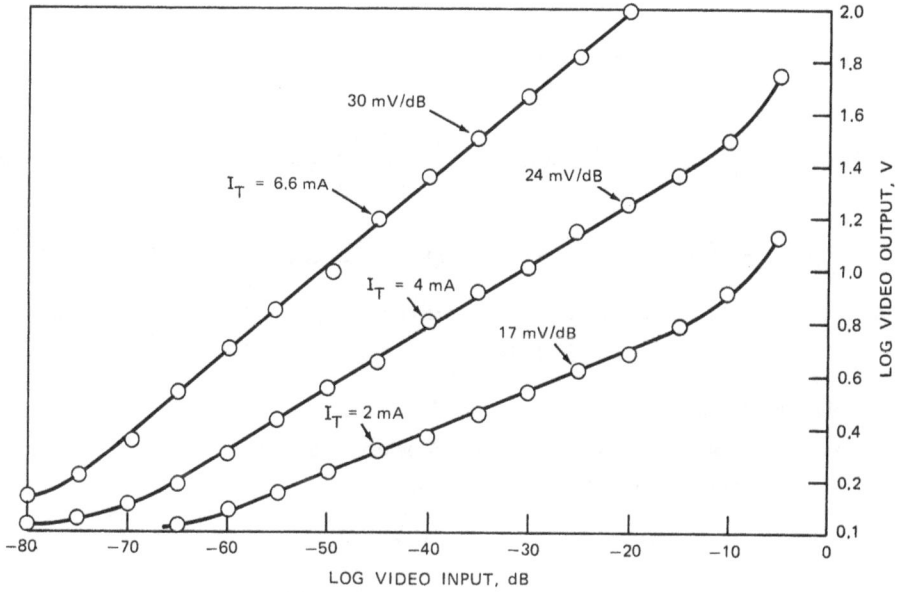

FIGURE 2-40. Three-Stage Series Linear-Limiting Transfer Function.

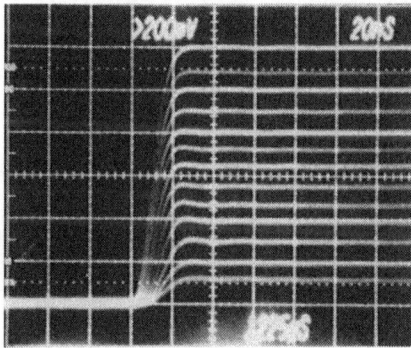

FIGURE 2-41. Pulse Response for
the Three-Stage Series, Lin-Limit
Logarithmic Video. Inputs from
-10 to -70 dB in 5-dB Steps
(I_T = 4 mA).

References

1. Risley, A. R., "Designers Guide to Logarithmic Amplifiers," *EDN*, 5 August 1973, pp. 42–51.

2. Sheingold, D. and F. Pouliot, "The Hows and Whys of Log Amps," *Electronic Design*, 1 February 1974, pp. 52–59.

3. Hughes, R. S., *Logarithmic Video Amplifiers*, Dedham: Artech House, 1971.

4. Solms, S. J., "Logarithmic Amplifier Design," *IRE Transactions on Instrumentation*, December 1959, pp. 91–96.

5. Stanford Electronics Laboratories. A Wideband Limiting-Summation Logarithmic Video Amplifier Design, by W. R. Kincheloe, Systems Techniques Laboratory, 6 June 1960. (Technical Report No. 560-1.)

6. Interdesign Co. Design Manual - Linear Monochip.

7. Exar Co. Design Manaul – Master Chip Custom Design Manual.

8. Linear Technology, Inc. Design Manual – Semicustom Array Design Manual.

9. Hughes, R. S., "Tunnel Diodes Excel as DC-Coupled Detectors," *Microwaves*, June 1981, pp. 59–62.

10. Tatum, J. and K. Hinton, "Tunnel Diodes Complement High-Performance Detectors," *Microwaves and RF*, February 1985, pp. 115–124.

11. Hughes, R. S., "Practical Approach Makes T_{ss} Measurement Simple," *Microwaves*, January 1982, pp. 69–72.

12. Clifford, D., "Approximate True Log Output at High Frequencies," *Electronics*, 31 January 1972, pp. 70–72.

13. Potson, D. and R. S. Hughes, "DC-Coupled Video Log Amp Processes 10 nsec Pulses," *Microwaves and RF*, April 1985, pp. 85-90, 150 and May 1985, pp. 75-78, 278-280.

14. Barber, W. L. and E. R. Brown, "A True Logarithmic Amplifier for Radar IF Applications," *IEEE Journal of Solid-State Circuits. Vol. SC-15 No. 3*, June 1980, pp. 291-295.

Chapter 3

LOGARITHMIC INTERMEDIATE FREQUENCY (IF) AMPLIFIERS

The input frequency and bandwidth of detector-logarithmic video amplifiers are limited only by the detector. Thus, RF in excess of 18 GHz and bandwidths in excess of 15 GHz are available. However, the instantaneous input dynamic range for detector-logarithmic video amplifiers is limited* (depending on detector) to some 55 dB (-45 to +10 dBm). Logarithmic IF amplifiers with an instantaneous dynamic range (IDR) in excess of 85 dB (depending on the IF and bandwidth) are commercially available. Modern units have excellent pulse response and work equally well for pulses or continuous waves (CW) (a situation that can be handled only with "true" DC-coupled logarithmic videos).

To design and fabricate high-frequency logarithmic IF amplifiers, access to high-frequency test equipment, computer simulation and optimization programs, and a hybrid design/construction facility is required that may not be available to the general designer. This chapter discusses various logarithmic IF amplifiers that are commercially available; however, the basics of each technique are presented.

Successive Detection Logarithmic IF Amplifiers

The logarithmic IF amplifier is an ancient (by modern standards) concept with the early tube versions having IF generally less than 60 MHz [1]. Early transistor designs [2, 3], patterned closely after the earlier tube designs, were generally limited to frequencies below 60 MHz. Figure 3-1 illustrates the basic block diagram for a typical successive detection logarithmic IF amplifier. The term *successive detection*

---

*The use of a high-frequency preamplifier, power divider, and a second detector-logarithmic video amplifier will extend the dynamic range, (see Figure 1-6); however, this is fairly costly and compromises the upper frequency limit.

derives from the fact that each detector (or logarithmic element (see Figure 2-18)) successively contributes to the output. Linear amplifiers A_1 to A_n amplify the input signal and drive detectors D_2 to D_n. Detector D_n detects the lowest level IF signal, supplied from A_n, and the video signal drives the video summing amplifier as shown. As A_n limits, detector D_{n-1} supplies the video signal, via a delay line, to the summing amplifier. This process continues until all linear stages are limited, at which time the input contributes directly, via D_1, to the video output. This process is illustrated in Figure 3-2 (see Figure 2-19). The delay lines ensure that the video signals arrive at the video amplifier simultaneously and greatly improve the video pulse response. This concept has been used in detector–logarithmic videos for the same reasons [4].

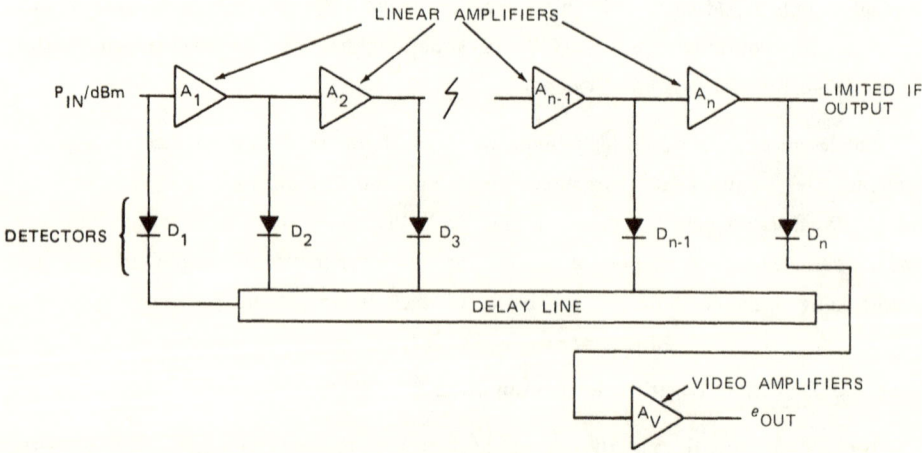

FIGURE 3-1. Successive Detection Logarithmic IF.

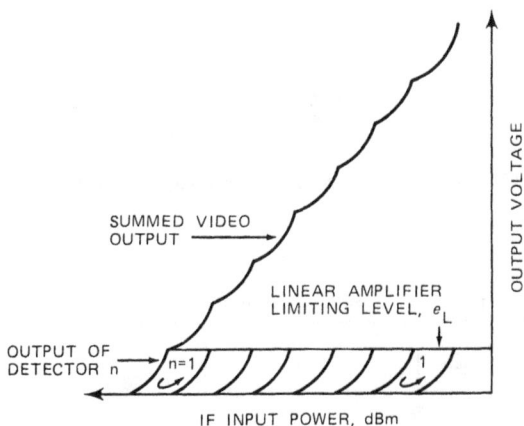

FIGURE 3-2. Successive Detection Logarithmic IF Characteristics.

Logarithmic IF amplifiers of the 1950–60s tend to be quite complex (and physically large) objects. Each linear stage required IF tuning to obtain the necessary IF bandwidth. Designing and tuning a logarithmic IF tended to be most time-consuming, and more often than not the designer faced the agony of defeat. That reliable devices were marketed is a true testimony to the talent and tenacity of those involved. Frequencies and bandwidths increased with faster transistors; however, a major improvement, both with respect to performance and physical size, occurred when Lamagna [5] avoided the need for video delay lines by using the wideband, differential amplifier–detector configuration. The basic block diagram of this new technique is illustrated in Figure 3-3, with Figure 3-4 showing the basic differential amplifier–detector stage. As the IF signal increases at the base of the IF amplifier, Q_1, the detector, Q_2, turns off, thus decreasing its collector current and increasing the video voltage developed across R_L. By proper control of the current ratio of the IF (Q_1) and detector (Q_2) transistors (R_{e_1} and R_{e_2}), a logarithmic amplifier can be designed that will have excellent pulse response and will be insensitive to IF variations. Figure 3-5 illustrates the pulse response for a commercial logarithmic IF employing this concept (RHG model ICLT-300A) with the transfer function illustrated in Figure 3-6. Note the pulse response is excellent, as is the recovery time (more on this topic in Chapter 5). The logarithmic conformity (error) is within ±1 1/2 dB for input frequencies from 200 to 400 MHz. This technique is usable for IF greater than 1000 MHz [6].

59

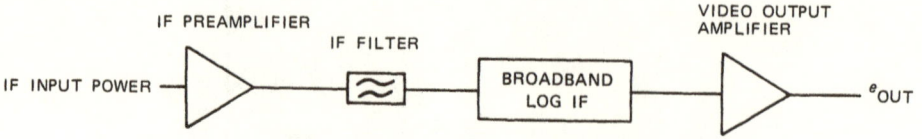

FIGURE 3-3. Basic Broadband Logarithmic IF Amplifier Concept.

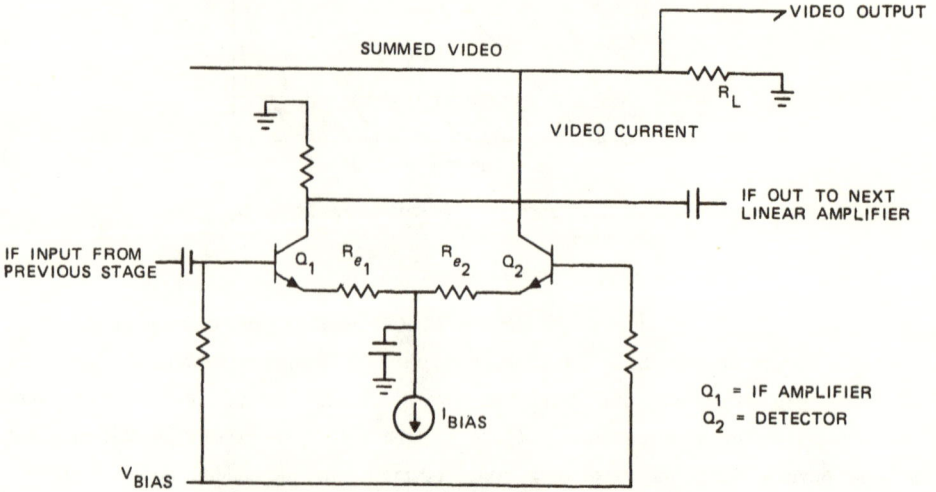

FIGURE 3-4. Basic Differential Amplifier-Detector Stage.

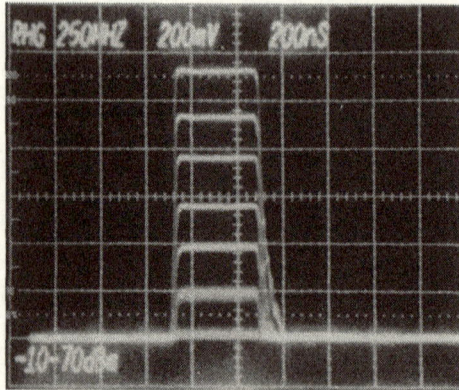

FIGURE 3-5. Pulse Response for a Differential
Amplifier-Detector Logarithmic IF (RHG
Model ICLT-300A) Inputs from –10 to
–70 dBm in 10–dBm Steps. IF = 250 MHz.

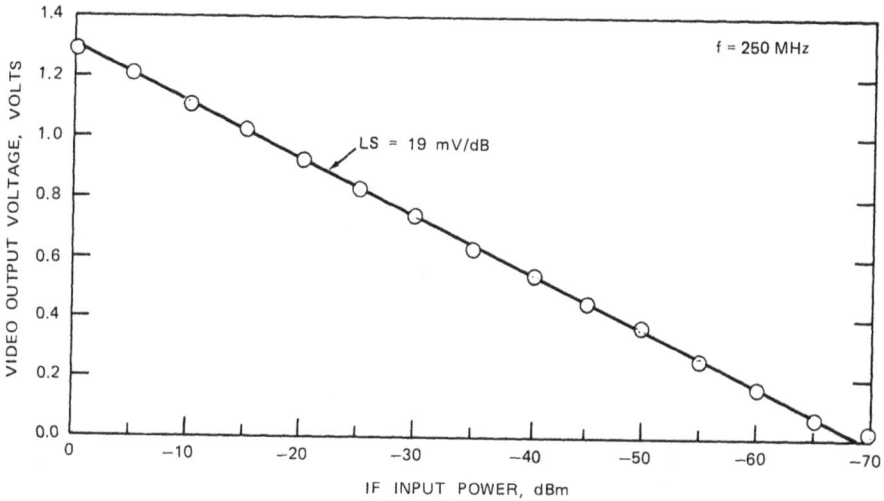

FIGURE 3-6. RHG Model ICLT 300-A Video Output Versus IF Power.

An integrated circuit of a modified version of this technique was produced by Plessey Semiconductors [7, 8]. Figure 3-7 illustrates the basic approach. A differential amplifier (Q_1, Q_2, and Q_3) with alternating current (AC) feedback (R_2 and R_1) provides a symmetrically limited high-frequency 12-dB gain (see Appendix E for a general description of this configuration). Transistors Q_4 and Q_5 form the detector. Transistor Q_5 is biased (V_B) at a voltage slightly lower than the base of Q_4; thus no collector current flows through Q_5. When an IF signal is present on the base of Q_4, Q_4 turns off, turning Q_5 on (causing collector current to flow to Q_5). The emitter resistor, R_e, causes the video collector current in Q_5 to gradually increase with increasing IF input over the 12-dB range of operation of the high-frequency linear amplifier (Q_1–Q_3). Thus, a logarithmic output characteristic can be obtained over the 12-dB range of the linear amplifier. The individual stages can be cascaded with the collectors of Q_5 tied together. Due to the wide bandwidth of the linear amplifier, no video delay lines are necessary. The Plessey Semiconductor SL 1521 is an excellent example of this concept [see 8], and IF up to 200 MHz are possible; however, the logarithmic conformity is not as good (±3 dB) as the designs presented previously.

61

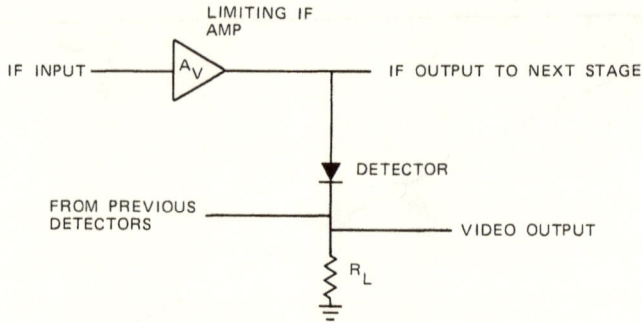

LIMITING IF
AMP

IF INPUT —— A_V —— IF OUTPUT TO NEXT STAGE

DETECTOR

FROM PREVIOUS
DETECTORS

VIDEO OUTPUT

R_L

(a) Basic approach.

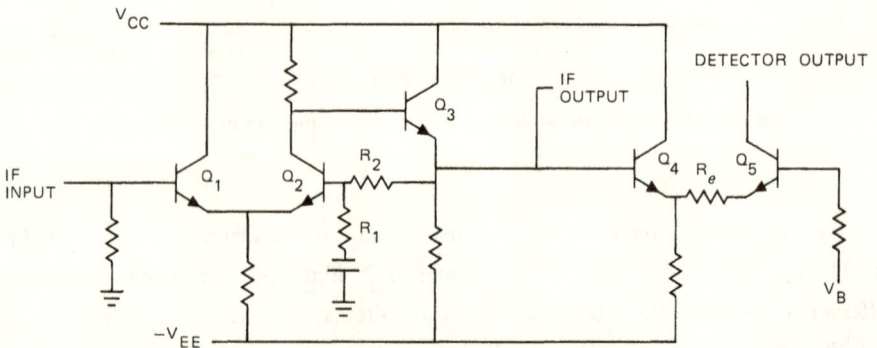

V_{CC}

Q_3

IF
INPUT

Q_1 Q_2 R_2

R_1

IF
OUTPUT

DETECTOR OUTPUT

Q_4 R_e Q_5

V_B

$-V_{EE}$

(b) Basic schematic.

FIGURE 3-7. Integrated Circuit Sequential Detection Logarithmic IF Stage.

Lansdowne and Kelly present a unique variation on the sequential summation logarithmic IF uses a video limiter following the detector [9] in conjunction with wideband limiting IF amplifiers as illustrated in Figure 3-8. The advantages of video limiting are

1. Increasing IF and bandwidth since video limiting occurs while the linear amplifiers are still in their linear region, thus limiting distortion is eliminated

2. The video limiter can be designed to be temperature-insensitive and much more accurate than the conventional linear amplifier–detector approach

3. Varying the limiting level independently of the linear amplifier, thus permitting accurate control of the logarithmic characteristic

4. Eliminating the video delay lines

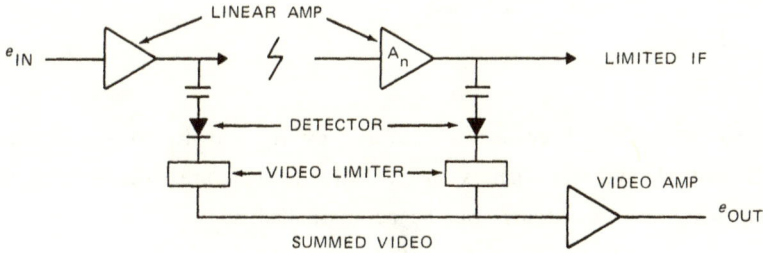

FIGURE 3-8. Logarithmic IF Using Video Limiters.

A simplified linear stage is illustrated in Figure 3-9 [10]. The voltage gain is dependent upon the effective AC collector resistance r_c, R_e, and the transistor dynamic emitter resistance r_e'. The dynamic emitter resistance can be given as

$$r_e' \cong \frac{86.25 \times 10^{-6}\ T}{I_E}\ \Omega \tag{3-1}$$

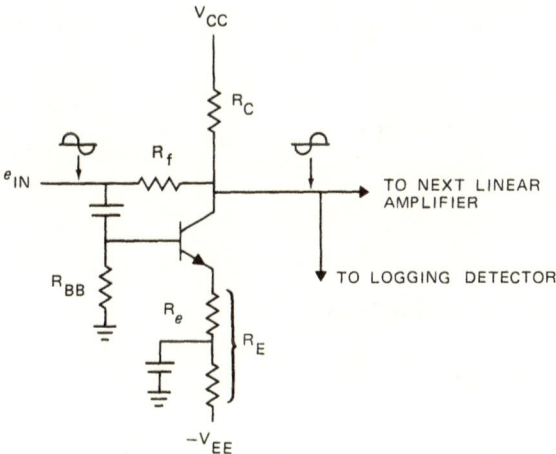

FIGURE 3-9. Basic High-Frequency-Limiting Amplifier.

63

where

T = temperature in degrees K

I_E = transistor emitter bias current

Equation 3-1 reduces to the well-known equation

$$r_e' = \frac{0.026}{I_E} \ \Omega \tag{3-2}$$

for $T = 300°K$ (27°C). Feedback is provided by R_e and R_f to ensure the gain is reasonably independent to changes in r_e' with temperature. The output limit for negative inputs (positive outputs) can be approximated as

$$e_o\bigg|^{+}_{max} \cong I_E \ r_c \tag{3-3}$$

where

$$e_o\bigg|^{+}_{max}$$ = maximum positive output

r_c = effective AC collector resistance

The designer must ensure the stage does not saturate for positive inputs (negative outputs). The saturation output for negative outputs may be approximated as

$$e_o\bigg|^{-}_{max} = \frac{A_V}{A_V + 1} \ (V_C - V_B) \tag{3-4}$$

where

$$e_o\bigg|^{-}_{max}$$ = maximum negative output

V_C = DC collector voltage

V_B = DC base voltage

64

Thus, to ensure nonsaturation, the input must not exceed

$$\left. e_{in} \right|_{max}^{+} \leq \frac{\left. e_o \right|_{max}^{-}}{A_V} \tag{3-5}$$

or

$$\left. e_{in} \right|_{max}^{+} \leq \frac{V_C - V_B}{A_V + 1} \tag{3-6}$$

Since $\left. e_{in} \right|_{max}^{+}$ is the output of the previous amplifier stage, to avoid saturation substitute Equation 3-3 into Equation 3-6

$$I_E \, r_c \leq \frac{V_C - V_B}{A_V + 1} \text{ (to avoid saturation)} \tag{3-7}$$

To illustrate the above discussion, let

$$I_E = 5 \text{ mA}, \ r_c = 80, \ I_E = 5 \text{ mA} \ (r_e{}' = 5.2\Omega)$$

$$A_V = 3.16 \ (10 \text{ dB}), \ V_C = 6V, \ V_B = 0, \ R_e = 20\Omega$$

$$\left. e_o \right|_{max}^{+} = (5 \text{ mA})(80) = 0.4V \text{ (Equation 3-3)} \tag{3-8}$$

$$\left. e_o \right|_{max}^{-} = \frac{3.16}{4.16} \ (6) = 4.56V \text{ (Equation 3-4)} \tag{3-9}$$

and

$$\left. e_{in} \right|_{max}^{+} \leq \frac{6}{4.16} = 1.44V \text{ (Equation 3-6)} \tag{3-10}$$

65

and, since the maximum positive input from the preceding stage is 0.4 V, saturation is not a problem. One word of caution, however, the designer must ensure that the negative signal input is not sufficiently large to exceed the emitter-base breakdown voltage of the transistor.

Figure 3-10 illustrates the pulse response for a commercial logarithmic IF employing this video limit technique (Anzac, Model AM-350). As with the differential amplifier-detector of Figure 3-4, the pulse response is excellent with a recovery time of less than 200 nsec. The logarithmic transfer function is illustrated in Figure 3-11, and, as can be seen, the logarithmic conformity is less than ±1 1/2 dB. IF limiting is a by-product of successive detection and is also shown in Figure 3-11 (this signal can be detected and used as a signal threshold). Figure 3-12 shows the video output versus frequency (100 to 300 MHz) and, as can be seen, the output is flat to less than ±2 dB for inputs from +5 to -70 dBm. Commercial units using this design concept can be obtained for frequencies in excess of 1.5 GHz from several companies [11, 12].

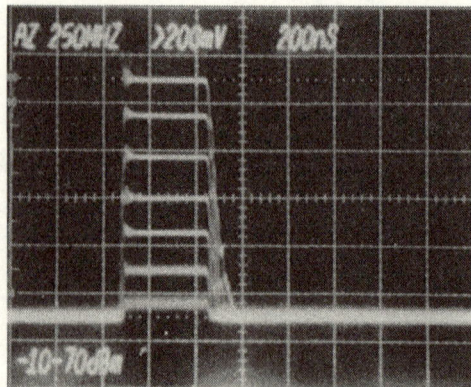

FIGURE 3-10. Pulse Response for Eight-Stage Detector-Video Limiter Logarithmic IF (Anzac Model 350) Inputs From -10 to -70 dBm in 10-dB Steps, IF = 250 MHz.

66

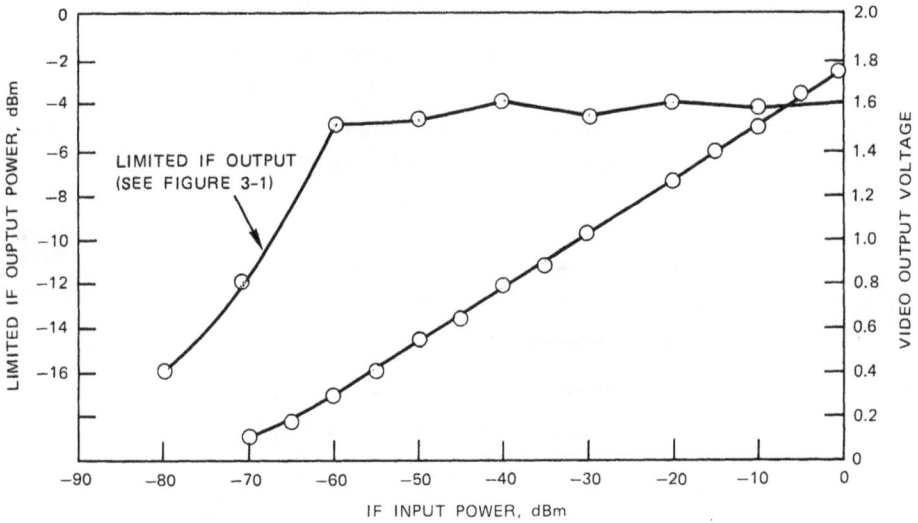

FIGURE 3-11. Video Output Versus IF Input Power Anzac Model AM-350, Eight-Stage, IF = 250 MHz.

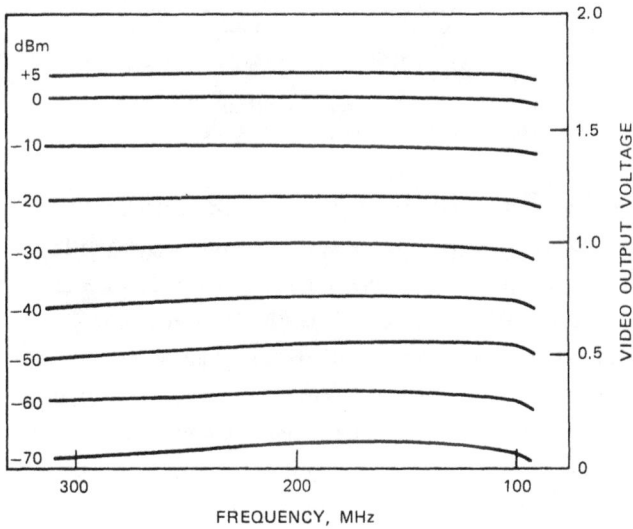

FIGURE 3-12. Video Output Voltage Versus Intermediate Frequency. Anzac Model AM-350 [11].

67

The reader may well ask, "What is the video output characteristic with frequency for the detector-logarithmic video amplifier discussed in Chapter 2 (Figure 2-35)?" Figure 3-13 illustrates the video output for IF from 250 MHz to 2 GHz. As can be seen, the flatness is less than ±1 dB over this frequency. This flatness is one of the strong points of the detector-logarithmic video amplifier.

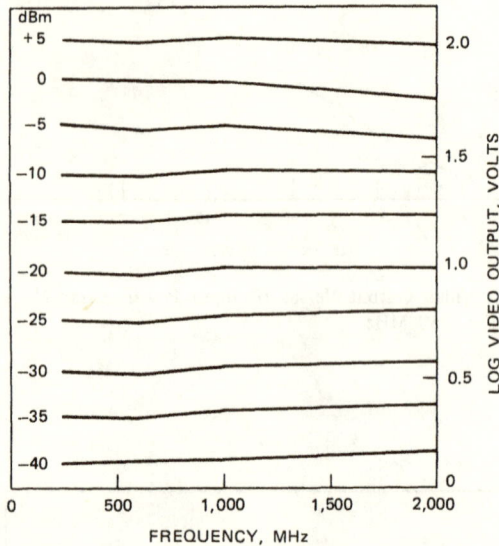

FIGURE 3-13. Detector-Logarithmic Video Output Versus Frequency (see Figure 2-35).

The instantaneous dynamic range of successive detection logarithmic amplifiers exceeds that of detector-logarithmic videos. However, the logarithmic IF has another primary advantage. It gives the same output voltage for CW as for pulses. This requires true DC coupling for detector-logarithmic videos (Chapter 4); however, it is inherent in well-designed logarithmic IF because all gains are AC-coupled. Thus, any DC drift in the linear IF biasing is not coupled to the output. The temperature stabilization of detectors, video limiters, and output video amplifier is a *much* simpler matter than designing a true DC-coupled logarithmic video (Chapter 4). This inherent CW capability can, however, be a disadvantage if one is interested in receiving pulses in a CW environment (termed CW rejection). This is discussed in Appendix H.

True Logarithmic IF Amplifier

The definition of true logarithmic IF amplification has caused some confusion. The term true in this context refers to the fact that the *IF* output voltage is a logarithmic function of the IF input power. Receiver designers can thus perform their magic at the IF level (most useful in MTI receivers) over a wide instantaneous input dynamic range, prior to detection [13-15].

The series lin-limit logarithmic video amplifier discussed in Chapter 2 (Figures 2-38 and 2-39), due to the inherent dual-polarity limiting of the differential amplifier, lends itself nicely to true logarithmic IF applications. Commercial integrated circuit versions are available for IF up to 200 MHz [see 8] with discrete designs capable of operating beyond 1 GHz [see 14]. Barber and Brown [see 15] describe the operation of an 80-dB dynamic range, 70 MHz true logarithmic amplifier with the following characteristics:*

Instantaneous dynamic range, dB 80

Logarithmic conformity, dB .... ±1

Intermediate frequency, MHz 70

Low level composite gain $(A_V + 1)$ .... 3.16 (10 dB)

Stage-limiting level, e_L, Vrms 0.1

Logarithmic-slope $\dfrac{e_L}{20 \text{ Log } (A_V + 1)}$, mV/dB 8.1

Stages 8

Output .... IF (no detector)

Integrated circuit, Plessey .... SL 531

Figure 3-14 illustrates the IF output, in root-mean-square (rms) volts versus IF dBm input. The measured LS is 9 mV/dB with a logarithmic error less than ±1 dB [see 8].

This logarithmic technique, whether video or IF, has an inherently low phase shift versus input power variations. Figure 3-15 illustrates the IF output phase variation as a function of input intensity.

---

*The schematic diagram is essentially that shown in Figure 2-39.

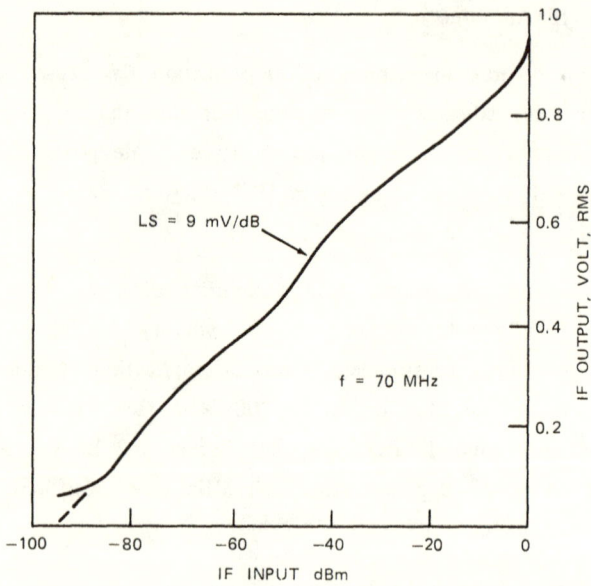

FIGURE 3-14. True Logarithmic IF Transfer Function [8].

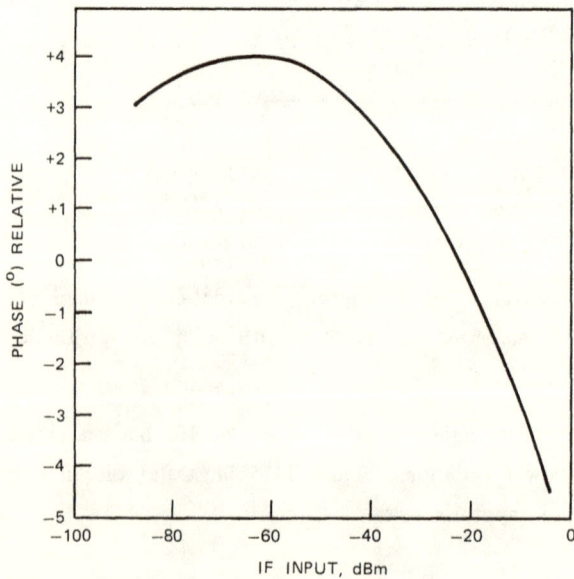

FIGURE 3-15. Phase Variation Versus Input Power [8].

70

A linear detector (Appendix F) is generally used with true logarithmic IF amplifiers since the IF output is, in most practical designs, above −10 dBm at the start of logarithmic action. With reference to Figure 3-14, logarithmic action starts $P_{in}\Big|_t^{dBm}$ at −90 dBm with a rms output of 0.1 V, or $e_{out}\Big|_t^{dBm}$ = 20 log rms + 13 = −7 dBm, and this is well within the linear region for a Schottky detector (Appendix F).

To give the reader an idea of what the true log IF output looks like, a sine wave input was provided to the series lin–limit logarithmic video of Figure 2-39 (this is, in reality, a true logarithmic IF). Figure 3-16 illustrates the output (logarithmic IF). The low phase shift is inherent with this technique.

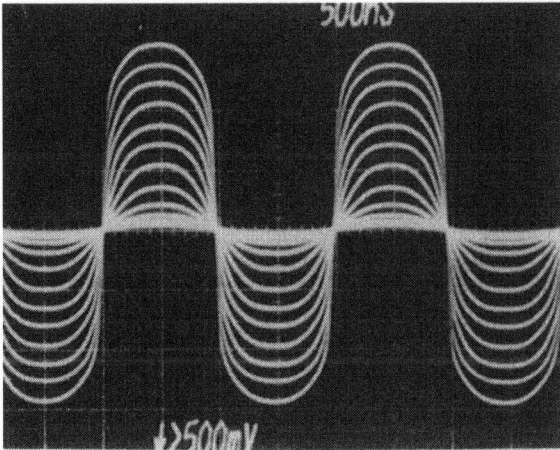

FIGURE 3-16. True Logarithmic IF Output (see Figure 2-39 for circuit diagram). Input Voltage From −20 to −70 dBV in 5 dB Steps.

Logarithmic IF amplifiers have progressed rapidly in the areas of frequency, bandwidth, and size during the past 10 years. Frequency has been pushed up to 2 GHz, with IF bandwidths exceeding 500 MHz, and thin film technology has decreased the size to 2.5 by 1 by 0.25 inch (0.625 in^3). The gain stages are AC-coupled, thus greatly easing drift problems due to temperature. Pulse response is exceptional, and pulse widths less than 20 nsec can be handled easily. So why the need for detector-logarithmic video amplifiers?

1. Logarithmic conformity over wide bandwidths is far superior for detector-logarithmic videos.

2. Output tracking for matched units over a wide bandwidth is superior for the detector-logarithmic video.

3. CW rejection (see Appendix H) is most difficult for logarithmic IF amplifiers.

4. Design and construction for high-frequency logarithmic IF amplifiers requires expensive test equipment, etc.; thus commercial units are usually required. The design and construction of detector-logarithmic video amplifiers is much simpler.

5. Logarithmic IF amplifiers tend to get expensive as the IF and bandwidth are increased ($1500 for a 750-MHz unit and $2500 for a 1.8-GHz unit). Commercial detector-logarithmic videos are approximately one-half this cost.

References

1. Chambers, T. H. and I. H. Page, "The High-Accuracy Logarithmic Receiver," *Proceedings of the IRE*, August 1954, pp. 1307-1314.

2. Alcock, R. N., "A Wide Band Transistor Logarithmic Amplifier at 45 Mc/s," *Electronic Engineering*, July 1962, pp. 444-449.

3. Amperex Electronic Corp., *A Transistorized Wide-Band Logarithmic Amplifier at 45 Mc Using the Amperex 2N2084.* (Report No. S-104. No date; however, schematic given is the same as in Reference 2 above.)

4. Hughes, R. S., *Logarithmic Video Amplifiers.* Dedham:Artech House, 1971.

5. Lamagna, J., "Design Compact Log Amps," *Microwaves*, April 1974, pp. 58-61.

6. RHG Electronics Laboratory, Inc., "Flatpack Log Amps are Merchants of Speed." *Microwaves and RF*, March 1984, p. 179.

7. Gay, M. S., "Log IF Strips Use Cascaded ICS," *Electronic Design*, 19 July 1966, pp. 56–59.

8. Plessey Semiconductors, *Broadband Amplifier Applications*. September 1984.

9. Lansdowne, K. and A. J. Kelly, "Microwave Logarithmic Amplifiers Using Hybrid Integrated–Circuits," Paper presented at the 1971 *IEEE International Solid–State Circuits Conference*. Digest of Technical Papers, 1971, pp. 94–95.

10. Watkins–Johnson. *Cascadable Amplifiers*, by D. L. Cheadle. (Tech–Notes, Vol. 6, No. 1, January/February 1979.

11. Anzac Product Catalog, 1982.

12. Varian Log Amplifier Catalog, Publication 4465, no date.

13. Waroncow, A. and J. Croney, "A True I.F. Logarithmic Amplifier Using Twin-Gain Stages," *The Radio and Electronic Engineer*, September 1966, pp. 149–155.

14. Loesch, B., "A UHF True Logarithmic Amplifier," *IEEE Transactions on Aerospace and Electronic Systems*, Vol. AES–9, No. 5, September 1973, pp. 660–664.

15. Barber, W. L. and E. R. Brown, "A True Logarithmic Amplifier for Radar IF Applications," *IEEE Journal of Solid–State Circuits*. Vol. SC–15, No. 3, June 1980, pp. 291–295.

Chapter 4

THE QUEST FOR THE TRUE DC–COUPLED DETECTOR–
LOGARITHMIC VIDEO AMPLIFIERS

The logarithmic IF amplifiers discussed in Chapter 3 have the same transfer function for either pulse or CW inputs. A true DC–coupled detector–logarithmic video has the same characteristic. As will be shown, DC coupling is difficult to achieve, and some form of DC nulling loop (or DC restoration) may be required. This is obviously not true DC coupling as occasionally the signal baseline must be restored.

This chapter presents the basic techniques and design procedures necessary to obtain true DC coupling followed by a simple DC nulling loop concept that is most useful with the detector–logarithmic video discussed in Chapter 2 (see Figure 2–35). It will be assumed that as much of the logarithmic video as practical will be implemented with a "semicustom" integrated approach [1–3], hence integrated circuit design techniques will be used [4–6].

Why DC Coupling?

A legitimate question. As will be seen, AC–coupled logarithmic videos place far fewer constraints on the linear video amplifiers driving the logarithmic stages [7]. The DC output voltage drift, with time and temperature of the linear video amplifiers, is the most critical parameter for DC–coupled logarithmic videos. So why DC coupling? Figure 4–1 illustrates the logarithmic response for the logarithmic amplifier discussed in Chapter 2 (see Figure 2–35) with logarithmic stages 4 and 5 captively coupled (0.01 μF) to the 30–dB video amplifier. The undershoot is obvious and expected. The top of the pulse is differentiated, and this may or may not be important (if the pulse is sampled close to the leading edge, as is often done in radar systems to minimize multipath, differentiation may be of limited concern). The primary problem is recovery time. When can another pulse be processed? With reference to Figure 4–1b, the recovery time, for a 1–μsec pulse, is 25 μsec (or a pulse repetition frequency (PRF) of 40 kHz). Many radar and EW systems require processing PRFs in excess of 100 kHz, so the simple AC–coupled logarithmic is not acceptable. If the processing of high density pulses *and* CW must be accomplished, DC coupling is the solution.

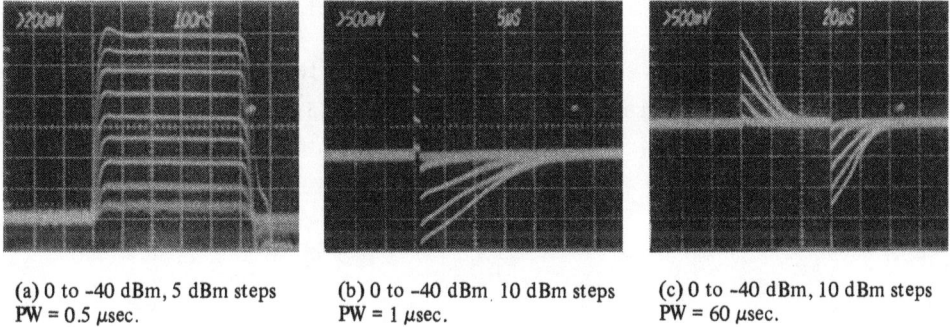

(a) 0 to –40 dBm, 5 dBm steps
PW = 0.5 μsec.

(b) 0 to –40 dBm 10 dBm steps
PW = 1 μsec.

(c) 0 to –40 dBm, 10 dBm steps
PW = 60 μsec.

FIGURE 4-1. AC Coupled Detector-Logarithmic Video Response (see Figure 2-35 for circuit diagram).

DC–Coupled Detector–Logarithmic Videos

Figure 4–2 illustrates the detector–logarithmic video configuration presented in Chapter 2. The discussion that follows is based on conclusions drawn from Appendix B.

Figure 4–3 illustrates a basic logarithmic stage with an attenuator and a current source. The attenuation is given by

$$\text{Atten} = \frac{R_2}{R_1 + R_2} \qquad (4\text{-}1)$$

or

$$\text{Atten} = \frac{1}{1 + R_1/R_2} \qquad (4\text{-}2)$$

75

Hence, the attenuation is a function of the ratio of R_1/R_2 and not the resistor's absolute value. This is most important since the tolerance for integrated resistors may be ±25%; the variation in resistor ratio from wafer to wafer will be 2% and only ±0.01%/°C with temperature [see 2].

The individual logarithmic stage bias current, I_T, is given as [see 4].

$$I_T = I \left(\frac{R_K}{R_E} \right) \qquad (4\text{-}3)$$

Thus this current is closely controlled by the ratio of two integrated resistors. The logarithmic error for a design using 15 dB/stage is ±0.45 dB (or ±0.225 dBm using a square law detector) for a current mismatch of 3% (Table A-3). This is well within reason for integrated transistors and resistors.

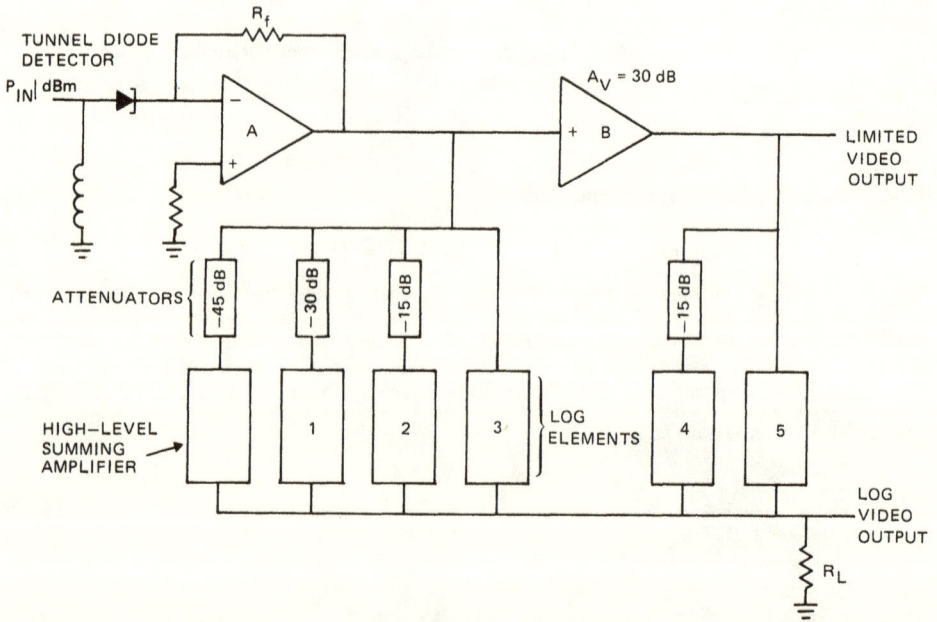

FIGURE 4-2. Basic Detector-Logarithmic Video Configuration.

76

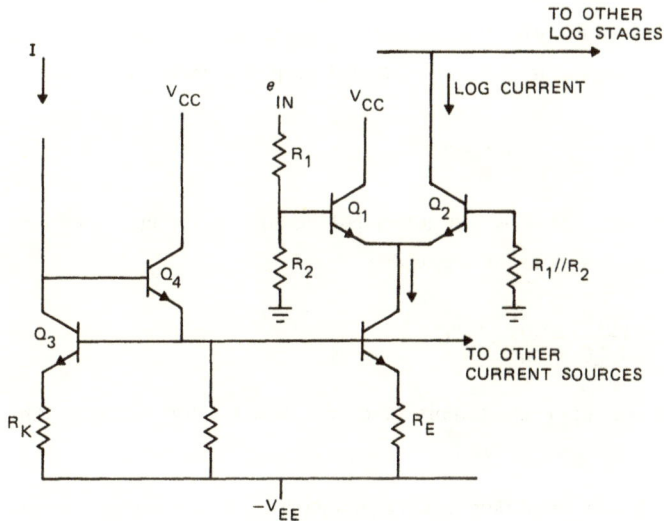

FIGURE 4-3. Typical Logarithmic Stage With Current
Bias (Q_3 and Q_4).

The logarithmic error is a strong function of any DC voltage appearing on the
bases of the logging transistors (Table B-2) and an offset voltage, ΔV, of 3 mV
(for 15 dB/stage) gives an error of ±0.86 dB (or ±0.43 dBm). Offset voltage error
due to the base bias currents can be minimized by ensuring that each base is driven
by the same, small value, resistance. Assuming the bases are at the same low poten-
tial, the base–emitter voltages for the logarithmic transistors are the same, and the
primary error contributed by the transistor is due to differences in the emitter
areas. Any error caused by emitter area mismatch will be temperature-insensitive.
This brings us to the linear video amplifiers. Using a ΔV of ±3 mV (this ensures
that two (or more) detector–logarithmic amplifiers will track to within ±0.43 dBm
(Table B-2)) places severe temperature and time drift requirements on the two
linear amplifiers. The gain of the second (limiting) linear amplifier in Figure 4-2 is
30 dB (31.62); thus this amplifier's input, for a 3–mV output, is

$$e_{in} = \frac{3mV}{31.62} \cong 95 \ \mu V \tag{4-4}$$

and this is the *maximum* output of the first amplifier. To put this in better perspective, assume that this offset occurs at a 100°C temperature and the offset is 0 μV at 25°C. The temperature drift of the first amplifier must be

$$\Delta e_0/\Delta T = \frac{95 \ \mu V}{75°C} \cong 1.3 \ \mu V/°C \tag{4-5}$$

which is tight for a high–speed integrated circuit operational amplifier. The drift of the second amplifier is more reasonable

$$\Delta e_0/\Delta T = \frac{3mV}{75°C} = 40 \ \mu V/°C \tag{4-6}$$

However, this amplifier must limit (and not saturate for recovery time reasons) for large inputs.

The two linear amplifiers have characteristics that existing commercial units do not provide. High–speed operational amplifiers (op–amp) can be used for the first amplifier; however, the 1.3 μV/°C is a real problem. A proposed solution to this problem [8] is the use of a DC–stable, low–frequency op–amp in conjunction with a high–speed op–amp as shown in Figure 4-4. This technique reduces the output voltage due to the temperature drift of the input offset voltage of the high–speed op–amp* to acceptable levels; however, this technique, alas, is not suited for logarithmic applications. The low–frequency op–amp senses the signal difference between the plus and minus inputs and causes an output response of its own. Potson [9] has studied this problem, and no simple solution has been found. Figure 4-5 illustrates the logarithmic response using the DC stabilizing approach of Figure 4-4 in Figure 2-36. The recovery time is inversely related to the pulse width, and the problem is obvious.

---

*This technique does nothing to improve the temperature effects of the offset currents.

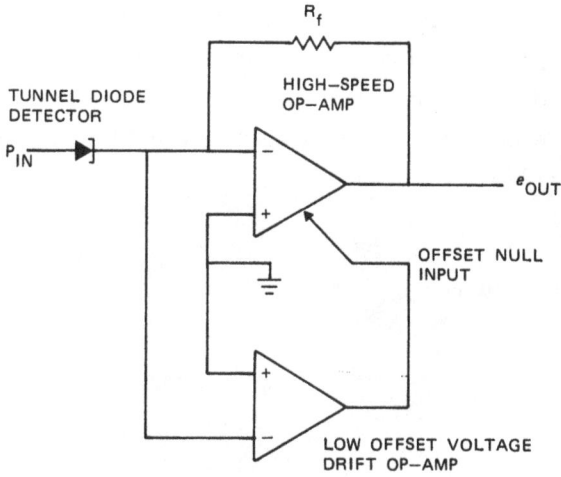

FIGURE 4-4. DC Stabilizing of a High-Speed
Op-Amp (not suitable for logarithmic
applications).

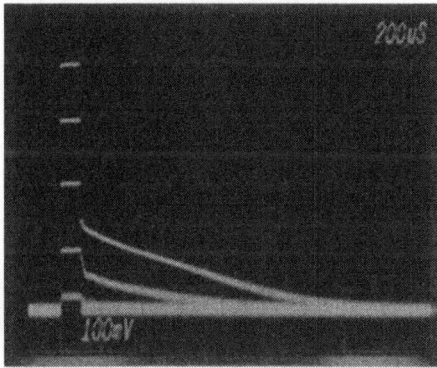

FIGURE 4-5. Recovery Time for Logarithmic
Amplifier Employing the Concept of
Figure 4-4. 0 to -40 dBm in 10-dBm steps
(I_T = 1 mA, R_L - 261 Ω).

79

A discrete design is probably the best solution.* MacPherson [10] describes the use of a supermatched pair of transistors in conjunction with a simple differential amplifier. This configuration is illustrated in Figure 4-6. Transistor Q_1 is a monolithic dual transistor (PMI, MAT 01) with the following pertinent characteristics (from $-55°C$ to $+125°C$):

maximum offset voltage, V_{OS} = 0.15 mV

maximum offset current, I_{OS} = 80nA

$\Delta V_{OS}/\Delta T$ = 0.5 $\mu V/°C$

$\Delta I_{OS}/\Delta T$ = 90 pA/°C

h_{FE} match = 7%

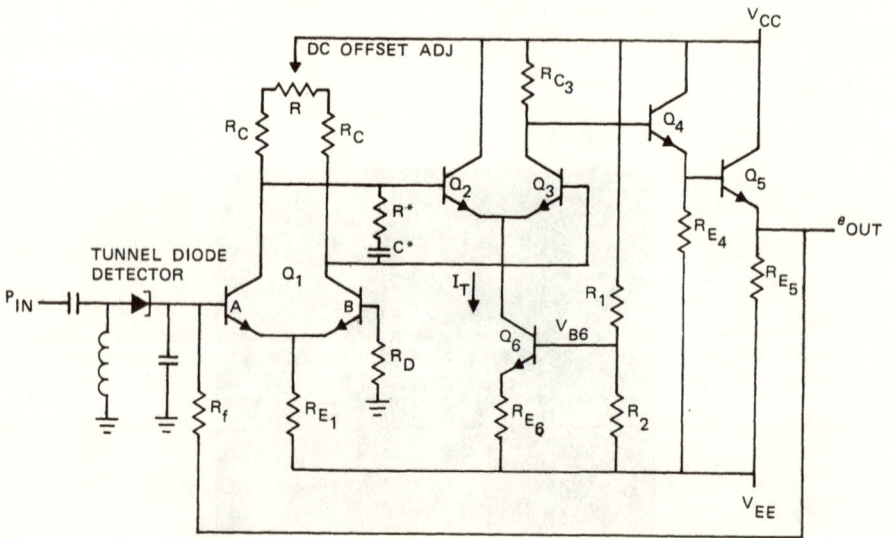

*Required for frequency stabilization.

FIGURE 4-6. DC Stable Video Amplifier Using Supermatched Transistors (Q_1A and Q_1B).

*The basic design of an integrated circuit op-amp for logarithmic video application is presented in Appendix E. This design has promise; however, the output offset voltage drift is excessive.

Transistors Q_2 through Q_6 are integrated circuit arrays that can be obtained from various semicustom integrated circuit suppliers [see 1-3].

The basic design equations are now given, with the following assumptions being made:

$$V_{BE_2} = V_{BE_3}, \; V_{BE_4} = V_{BE_5} = V_{BE_6}$$

For $V_O = 0$

$$V_{C_3} = 2 \, V_{BE} \tag{4-7}$$

and

$$V_{C_3} = V_{CC} - \frac{I_T \, R_{C_3}}{2} \tag{4-8}$$

$$V_o = V_{CC} - \frac{I_T \, R_{C_3}}{2} - 2V_{BE} = 0 \tag{4-9}$$

I_T is given as

$$I_T = \frac{V_{B_6} - V_{BE} - V_{EE}}{R_{E_6}} \tag{4-10}$$

and substituting this into Equation 4-9

$$V_o = V_{CC} - \frac{R_{C_3}}{R_{E_6}} \left(\frac{V_{B_6} - V_{BE} - V_{EE}}{2} \right) - 2 \, V_{BE} \tag{4-11}$$

or

$$V_o = V_{CC} - \frac{R_{C_3}}{R_{E_6}} \left(\frac{V_{B_6}}{2} + \frac{V_{EE}}{2} \right) + \frac{R_{C_3}}{R_{E_6}} \left(\frac{V_{BE}}{2} \right) - 2 \, V_{BE} \tag{4-12}$$

The terms V_{CC}, V_{EE} and V_{B_6} can be made temperature-insensitive. The values of integrated resistors do change with temperature; nevertheless, their ratios track quite well, i.e., within $\pm 0.01\%/°C$ [see 2].

To ensure V_O is temperature-insensitive, the right two terms in Equation 4-12 must cancel, or

$$\frac{R_{C3}}{R_{E6}} \left(\frac{V_{BE}}{2}\right) = 2V_{BE} \qquad (4\text{-}13)$$

Thus

$$\frac{R_{C3}}{R_{E6}} = 4 \qquad (4\text{-}14)$$

Substituting Equation 4-14 into 4-12 and solving for the value of V_{B6} to ensure $V_O = 0$, one obtains

$$V_{B6} = \frac{V_{CC}}{2} + V_{EE} \qquad (4\text{-}15)$$

V_{B6} is dependent upon V_{CC}, V_{EE}, and the ratio of R_1/R_2. This ratio may be found to be

$$R_1/R_2 = \frac{V_{B6} - V_{CC}}{V_{EE} - V_{B6}} \qquad (4\text{-}16)$$

The collector voltage of Q_3 is two V_{BE} drops above 0 V, so it makes sense to have its base 1 V_{BE} drop above 0 V. Therefore, the collector voltages of Q_1 (the supermatched dual) are 1 V_{BE} above ground. Thus, values for R_C and R_{E1} can now be determined.

Design example: $V_{CC} = +12$ V $V_{EE} = -6$ V
$$V_{BE} (Q_{4, 5, 6}) = 0.7 \text{ V}$$

$$V_{B6} \text{ (Equation 4-15)} = \frac{12}{2} - 6 = 0 \qquad (4\text{-}17)$$

Let $R_{C3} = 4.99$ k

$$R_{E6} = \frac{R_{C3}}{4} = 1.25 \text{ k} \qquad (4\text{-}18)$$

To ensure $V_{BE4} = V_{BE5} = V_{BE6}$, their emitter currents should be equal.

82

$$R_{E_4} = \frac{-V_{BE} - V_{EE}}{I_T} \tag{4-19}$$

$$R_{E_5} = \frac{-V_{EE}}{I_T} \tag{4-20}$$

where (Equation 4-10)

$$I_T = \frac{0 - 0.7 - (-6)}{1.25 \ k} = 4.24 \ mA \tag{4-21}$$

and

$$R_{E_4} = \frac{0.7 - (-6)}{4.24 \ mA} = 1.57 \ k\Omega \tag{4-22}$$

$$R_{E_5} = \frac{- (-6)}{4.24 \ mA} = 1.41 \ k\Omega \tag{4-23}$$

The parameter values for Q_1 are as follows.

Let R_C = 10 KΩ, R = 200 Ω, and V_C = 0.7 V

$$I_{C_A} = \frac{V_{CC} - V_C}{R_C} = I_{C_B} \tag{4-24}$$

or

$$I_{C_A} = \frac{12 - 0.7}{10.1 \ k\Omega} = 1.119 \ mA \tag{4-25}$$

Thus

$$I_{T_1} = 2 \ I_{C_A} = 2.238 \ mA \tag{4-26}$$

$$R_{E_1} = \frac{-V_{EE} - V_{BE}}{I_{T_1}} \tag{4-27}$$

or

$$R_{E_1} = \frac{- (-6) - 0.7}{2.238 \ mA} = 2.37 \ k\Omega \tag{4-28}$$

Figure 4-7 illustrates the complete amplifier, with the pulse response illustrated in Figure 4-8. The DC offset voltage is within ±1 mV for temperatures ranging from –40°C to +80°C.

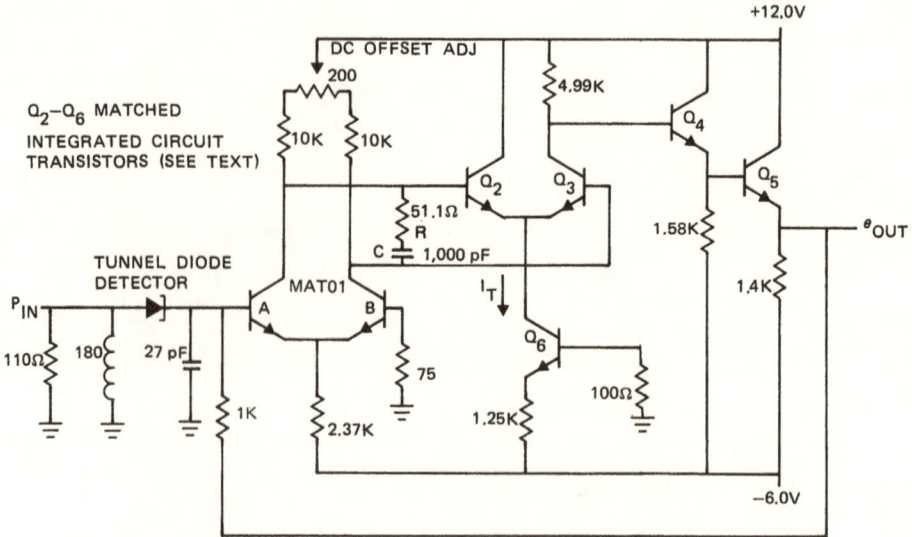

FIGURE 4-7. DC Stable Video Amplifier Using Supermatched Transistors.

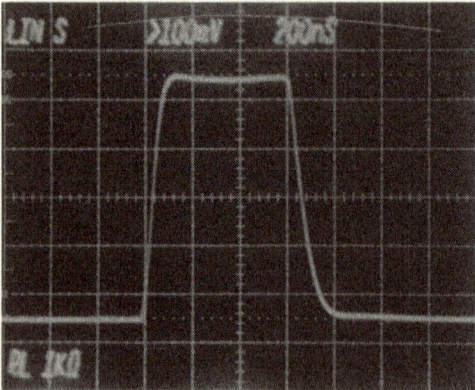

FIGURE 4-8. Pulse Response for Figure 4-7
(P_{in} = -10 dBm).

84

This configuration can be used, with modification, as the second lin–limit amplifier. The second amplifier will be overdriven by the first amplifier; thus, saturation* should be avoided. The output of the first amplifier will be some 5.5 V for a +5 dBm detector input; thus, the second amplifier can be driven into hard saturation. Figure 4-9 illustrates two methods to minimize this saturation. Diode D_1 limits the overdrive to transistor Q_1A. Diodes D_2 and D_3 limit the inputs to Q_2 and Q_3 to ±0.3 V. To ensure that the emitter–base voltage of the lower–level logarithmic stage is not exceeded, D_4 is needed. The limit voltage, V_L, can be obtained via a 5-V zener. Figure 4-10 illustrates the pulse response for this circuit. The DC output offset voltage was within ±3 mV from –40°C to +80°C.

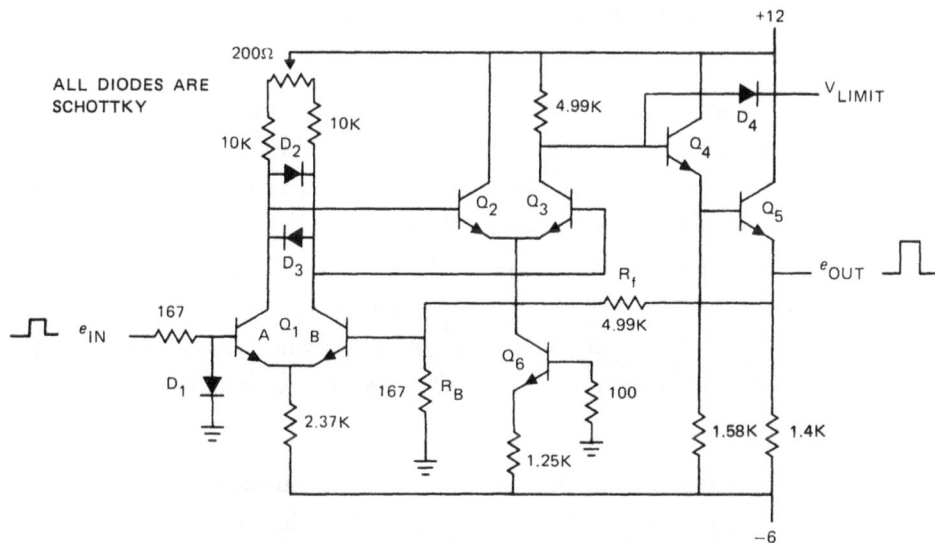

FIGURE 4-9. Non–Inverting Linear-Limit Video Amplifier (Voltage gain = 30 dB).

*Saturation is defined as the forward biasing of the base–collector junction and can cause excessive recovery time. Ideally, only a single differential amplifier stage should be used due to its inherent limiting (see Appendix E).

85

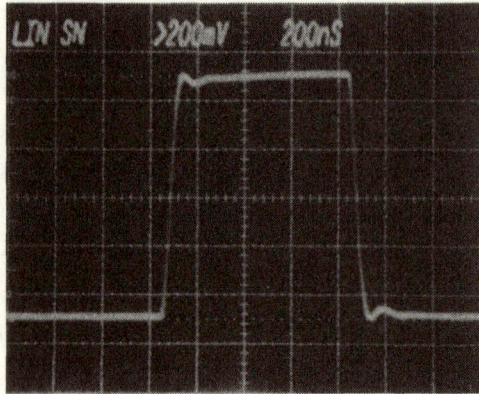

FIGURE 4-10. Pulse Response for the 30-dB
Linear Non-Inverting Amplifier (Figure 4-9).

The output stage is best accomplished via a common base transistor (discussed in Chapter 2) in conjunction with a DC level shifter. Figure 4-11 illustrates one such configuration (Appendix D). Transistor Q_1 is a common base amplifier that optimizes the pulse response (see Chapter 2). Transistors Q_2, Q_3, Q_4, and Q_5 comprise a DC level shifter (see Figure 2-35 and Appendix D). The LS can be set via R_L for a given application, and the output voltage can be set to zero via the DC output adjust. With respect to integration, all the components in Figure 4-11, except R_L and the DC output adjust, should be on the same integrated circuit as the logarithmic elements.

As shown, the design of a DC-coupled detector-logarithmic video amplifier is a touchy business. The linear amplifiers may have to be tested over temperature and then set for the best voltage offset. This is time-consuming and costly. Several firms market DC-coupled detector-logarithmic video amplifiers, and the reader should contact them as to specific requirements before beginning the quest for true DC coupling.

Do you really need true DC coupling? Many applications that at first glance suggest the need for this can use psuedo DC coupling.

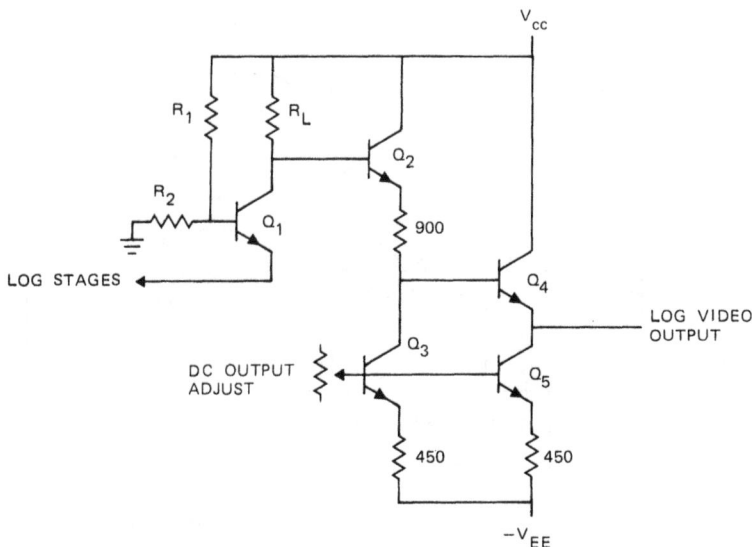

FIGURE 4-11. DC Logarithmic Video Output Stage.

Psuedo-DC Coupling

The primary difficulty with true DC coupling is the DC voltage drift of the linear amplifiers. The technique illustrated in Figure 4-12 greatly eases the linear amplifier voltage drift problem, and, as a by-product, provides inherent CW rejection (Appendix H). The switch, S, is normally closed, thus the output of A_{V_2} is driven (independently of any offset voltage) to 0 V via A_Δ and the integrator. The output voltage of A_{V_1} will be the offset voltage of A_{V_2} (if it were not driven to zero), V_{os2}, divided by the gain of A_{V_2}. For example:

$$V_{os2} = 20 \text{ mV (before DC nulling)} \qquad (4-29)$$

$$V_{o_1} = 20 \text{ mV}/31.6 = 0.63 \text{ mV} \qquad (4-30)$$

where V_{o_1} is the DC output voltage of A_{V_1}.

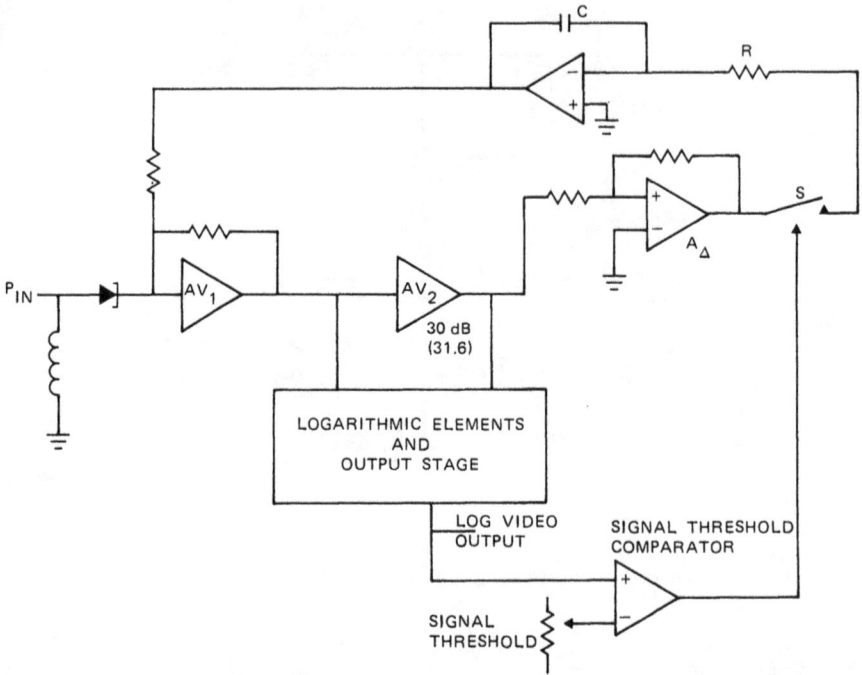

FIGURE 4-12. DC-Coupled Detector-Logarithmic Video.

When a pulse is present, the signal threshold comparator is triggered to open the switch (thus inhibiting the DC restoration loop from acting on the signal). This concept was used in conjunction with Figure 2-35, and the results are given in Figures 2-36 and 2-37. The DC loop dynamics can be controlled with A_Δ, R and C. There are many other possible configurations for this concept, limited only by the reader's needs and originality.

The circuit of Figure 4-12 will not work with CW; in fact, CW will be nulled out just like any DC offset voltage. Thus, the loop has inherent CW rejection (see Appendix H).

88

This chapter presented problems associated with true DC coupling with a start toward a realizable solution. People who sell true DC–coupled detector–logarithmic video amplifiers are continually improving their designs to obtain better DC and dynamic performance. If true DC coupling is not required and if one cannot tolerate the recovery time associated with AC coupling, (see Figure 4–1), the DC nulling loop may be the solution.

References

1. Interdesign Co., "Design Manual–Linear Monochip."*

2. Exar Co., "Design Manual–Masterchip Custom Design Manual."*

3. Linear Technology, Inc., "Semicustom Array Design Manual."*

4. Gray, P. R. & R. G. Meyer, *Analysis and Design of Analog Integrated Circuits.* New York; John Wiley & Sons, Inc., 1977.

5. Grebene, A. B., *Bipolar and MOS Analog Integrated Circuit Design.* New York: John Wiley & Sons, Inc., 1984.

6. Interdesign, Inc. *The Use of Bipolar Semiconductor Junctions in Linear Circuit Design*, by D. Bray. (Monochip Application Note APN–33.)

7. Hughes, R. S., *Logarithmic Video Amplifiers.* Dedham: Artech House, 1971.

8. Hughes, R. S., "Tunnel Diodes Excel as DC–Coupled Detectors," *Microwaves*, June, 1981, pp 59–62.

9. Potson, D., Comlinear Corp., private communication.

10. MacPherson, E., "A Semi–Custom IC for an Advanced ECM System," *Microwave Journal.* Sept 1981, pp 91–96.

*These companies will supply (for a fee) breadboard components such that your individual designs can be checked for DC performance.

Chapter 5

SPECIFYING LOGARITHMIC AMPLIFIERS FOR
RADAR AND EW APPLICATIONS

The previous chapters presented the basic types of logarithmic amplifiers suitable for use in radar and EW receivers (logarithmic IF amplifiers and detector-logarithmic video amplifiers) as well as various design implementations. The reader probably has visions of some type of logarithmic amplifier for her/his receiver. This chapter presents various factors in specifying the logarithmic amplifiers the reader should answer before any in-house design or logarithmic-amplifier purchase is made.

IF Center Frequency and Bandwidth

Logarithmic IF amplifiers can be obtained for frequencies up to some 2 GHz with an IF bandwidth of 500 MHz or so. These high-frequency units tend to get expensive. The change in logarithmic conformity (flatness) with frequency deteriorates as frequency and bandwidth increase. Detector-logarithmic video amplifiers are inherently broadband and quite flat with frequency.

Instantaneous Input Dynamic Range

Logarithmic IF amplifiers have a much larger instantaneous input dynamic range (-70 to $+10$ dBm) than detector-logarithmic video amplifiers, (-45 to $+5$ dBm). As the operating frequency and bandwidth increase, the input dynamic range for logarithmic IF amplifiers decreases.

Logarithmic Conformity (Maximum Logarithmic Error)

Figure 5-1 illustrates a typical logarithmic amplifier transfer characteristic. The logarithmic conformity over a specified bandwidth is usually better for detector-logarithmic video than for logarithmic IF amplifiers. If the bandwidth is greater than several hundred MHz, the logarithmic conformity for logarithmic IF amplifiers can easily exceed ±2 dB. Logarithmic conformity can be temperature–dependent with detector-logarithmic videos less dependent.

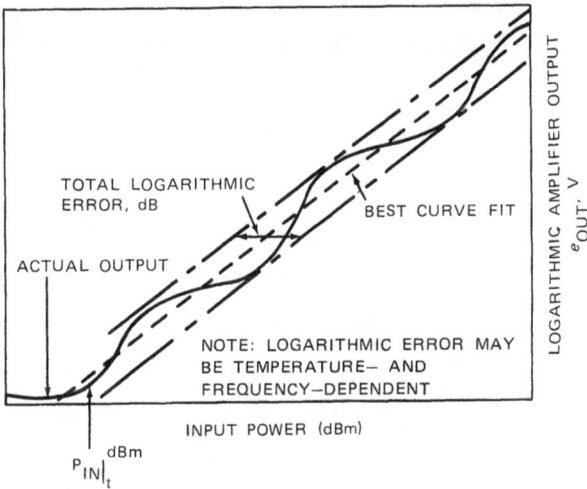

FIGURE 5-1. Logarithmic Amplifier Characteristics.

Linear-Logarithmic Transition Input Power, $P_{in}\big|_t^{dBm}$

The linear-logarithmic transition input power, $P_{in}\big|_t^{dBm}$, is illustrated in Figure 5-1. This level is much lower for logarithmic IF amplifiers (some –70 dBm) than for detector-logarithmic videos (–40 to –45 dBm). $P_{in}\big|_t^{dBm}$ may be temperature-sensitive for either type of logarithmic amplifier, and individual application will determine the allowable variation. The movement of $P_{in}\Big|_t^{dBm}$ with temperature for an uncompensated detector-logarithmic video (Appendix E) is 0.03 dB/°C (or a ±2 dB movement for a ±75°C temperature variation). Logarithmic IF amplifiers will have approximately the same variation.

Logarithmic Slope

The LS should be determined by system requirements. Too low a LS places severe requirements on post logarithmic amplifier circuitry. The LS for detector-logarithmic videos can easily be made quite temperature-insensitive. A variation of ±5 to ±10% should be expected for logarithmic IF amplifiers with temperature and frequency.

Figure 5-2 illustrates the effect of temperature variation on LS and $P_{in}\Big|_{dBm}^{e_o=0}$.

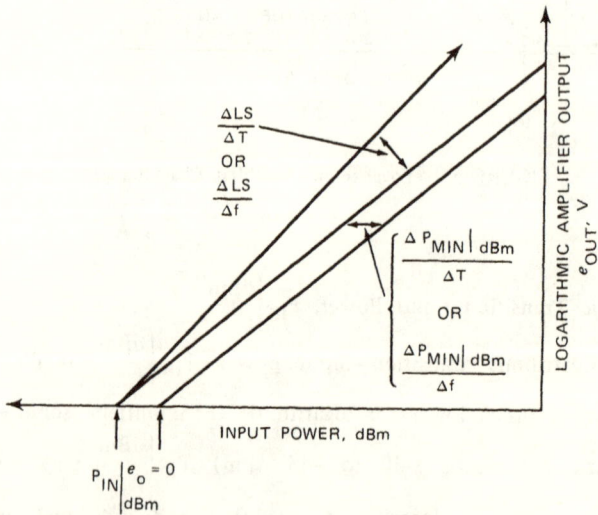

FIGURE 5-2. Basic Logarithmic Amplifier Temperature Characteristics.

The ideal logarithmic video output voltage is given as

$$e_{out}(\text{ideal}) = LS \left. P_{in} \right|_{dBm} - \left. P_{in} \right|_{dBm}^{e_o=0} \tag{5-1}$$

and generally system requirements will dictate the maximum permissible change in e_{out} with temperature. The designer then determines the necessary $\Delta LS/\Delta T$ and $\dfrac{\Delta P_{in} \Big|_{dBm}^{e_o=0}}{\Delta T}$ variations needed. It will be noted the change in $\left. P_{in} \right|_{dBm}^{e_o=0}$ and P_t dBm with temperature is the same.

Tangential Signal Sensitivity, $T_{ss} \big| dBm$

For logarithmic IF amplifiers $T_{ss} \big| dBm$ is much lower (~70 dBm) than for detector-logarithmic video amplifiers (-45 dBm). For a well-designed amplifier (IF or video), T_{ss} should be fairly insensitive to temperature variations.

Output rms Noise Level

If the logarithmic amplifier is to drive a signal threshold comparitor, any temperature dependence of the output rms noise voltage may require a change in the signal threshold voltage. The output rms noise voltage for a well-designed logarithmic IF or detector-logarithmic video amplifier should be insensitive to temperature variations.

Output DC Offset Voltage

The output DC offset voltage is the DC voltage present at the logarithmic amplifier output with no signal present. The temperature dependence on this voltage is most important if any change in offset voltage changes the output signal amplitude. Assume the following characteristics:

LS = 50 mV/dB
Output signal amplitude (DC offset = 0) = 200 mV
Output signal amplitude (DC offset = 100 mV) = 300 mV.

The 100–mV, DC offset voltage causes an effective 2–dB error (100 mV/ 50 mV/dB).

Selectivity

Selectivity and IF/RF bandwidth are usually synonymous for a linear receiver. The wide instantaneous input dynamic range of logarithmic amplifiers complicates the issue. This is best illustrated with a simple example:

Suppose a receiver has a logarithmic IF amplifier with an 85–dB input dynamic range (–80 to +5 dBm). A three–pole filter with an approximate attenuation of 60 dB/decade is placed in front of the log IF. Thus, for a frequency one decade (factor of 10) away from the filter's 3–dB frequency response, the filter attenuates the input signal by 60 dB. If the filter's input signal is –0 dBm, its output would be –60 dBm—well into the logging region of our logarithm IF amplifier. The net effect is that the effective receiver bandwidth, or selectivity, is greatly increased by using logarithmic amplification. Individual requirements dictate the number of filter poles necessary.

Video Rise Time, Video Bandwidth, and Minimum
Input Pulse Widths for a 1–dB Error

The output, or video, rise time is one of the most confusing parameter specifications associated with logarithmic amplifiers. This is because logarithmic amplifiers are nonlinear systems, and the common relationship between the 10 to 90% rise time $\tau_r \big|_{10\%}^{90\%}$ and 3–dB bandwidth is *not* valid

$$\tau_r \bigg|_{10\%}^{90\%} \neq \frac{0.35}{BW} \quad \text{(for logarithmic amplifier)} \tag{5-2}$$

A better and truer criterion of the pulse response for a logarithmic amplifier is the minimum input pulse width for a 1–dB error. Figure 5-3 illustrates the problem associated with specifying rise time. A logarithmic amplifier may have a 50–nsec rise time; however, it may take 125 nsec (measured from the 10% point) to settle

94

within ±1 dB of the final value (and the time may well be input intensity-dependent). Radar and EW designers are concerned with the minimum pulse width a system can handle and when (with respect to the leading edge) the received signal can be sampled for signal processing; consequently, the specification of the leading edge pulse characteristic is most important. This minimum input pulse width for a 1-dB settling error is a most reasonable specification. Decrease the input pulse width until the logarithmic output pulse has decreased by 1 dB, and measure this input pulse width (we will define the input (or linear) pulse width from the leading edge 50% point to the trailing edge 50% point). For pulse width or sampling times less than this, the signal will be in error in excess of 1 dB.

Figure 5-4 illustrates the logarithmic output for a wide pulse, logarithmic output for a pulse width giving a 1-dB error, and the input pulse width associated with the 1-dB error for the detector-logarithmic video amplifier illustrated in Figure 2-35. A digitizing oscilloscope was used for these pictures to integrate the noise. As can be seen, using the rise time as a criterion gives optimistic results. It may not be convenient to measure the input pulse width. A simple procedure that gives excellent results (if no pulse overshoot is present) is to measure the time from the 10% point to the 1-dB error value $\tau\Big|_{-10\%}^{-1\ \text{dB}}$. These values are given in Figure 5-4.

Recovery Time

Recovery time is defined as the time necessary to recover to a given specified level and is dependent on input intensity (and possibly input pulse width). One recovery time criterion is how soon a second pulse can be processed. Figure 5-5 shows how soon a second pulse can follow the first for a 1-dB amplitude error on the second (see Figure 2-35). The strong dependence of recovery time on input intensity is apparent. Well-designed logarithmic IF amplifiers will have a slightly shorter recovery time than detector-logarithmic videos. The dependence of recovery time for modern logarithmic IF amplifiers is usually independent of the input pulse width; however, the designer must be careful not to saturate any linear stages.

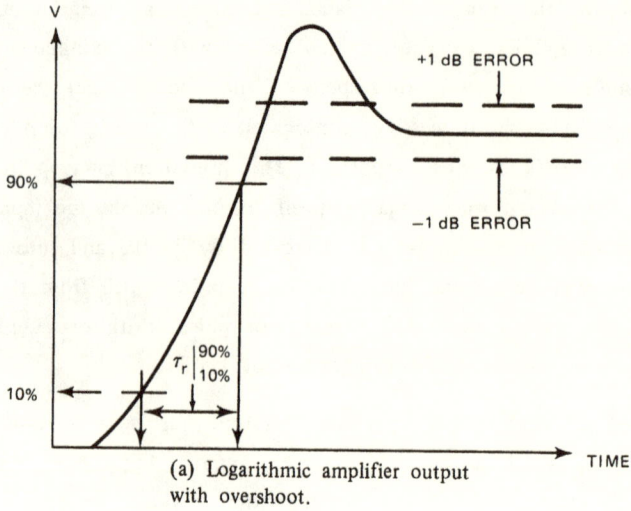

(a) Logarithmic amplifier output
with overshoot.

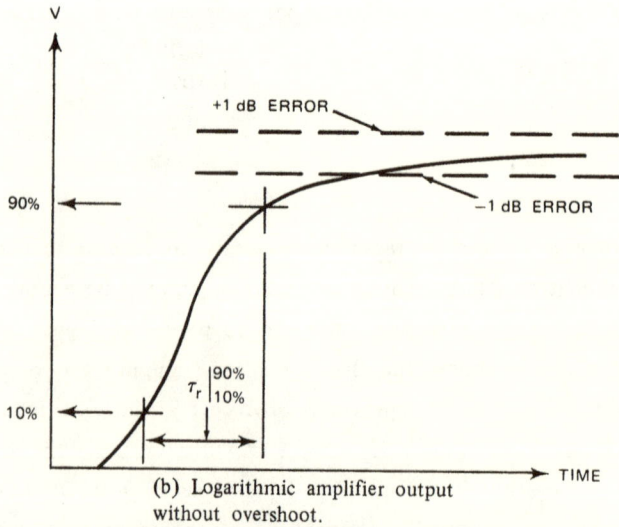

(b) Logarithmic amplifier output
without overshoot.

FIGURE 5-3. Logarithmic Amplifier Output Illustrating
Rise Time and 1-dBm Settling Error Time.

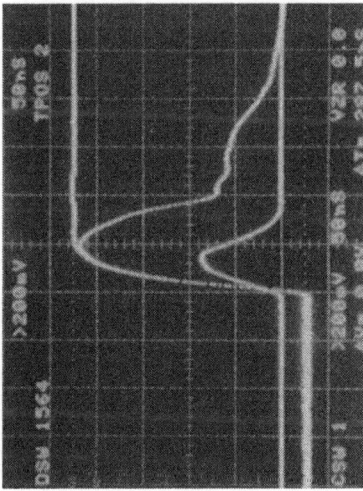

(a) $P_{in} = 0$ dBm.

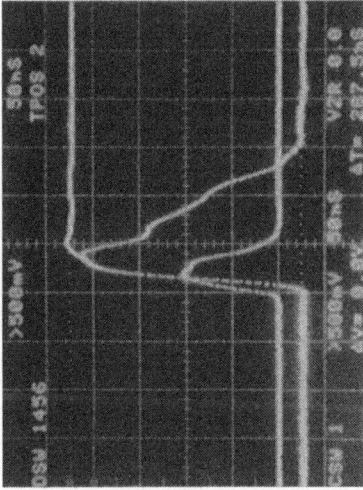

(b) $P_{in} = -20$ dBm.

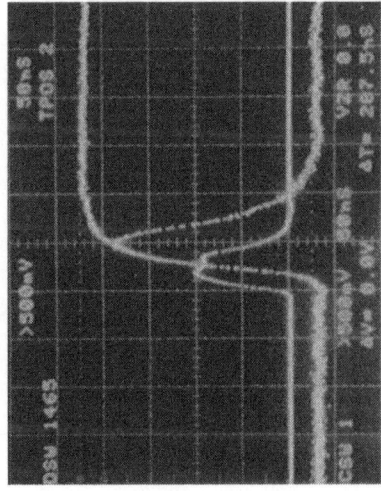

(c) $P_{in} = -35$ dBm.

P_{in}|dBm	τ_r|$^{90}_{10}$	PW	τ|$^{-1dB}_{10\%}$
0	30 nsec	50 nsec	48 nsec
-20	20 nsec	40 nsec	35 nsec
-35	40 nsec	40 nsec	40 nsec

FIGURE 5-4. Minimum 1 dB Pulse Width Detector-Logarithmic Video (see Figure 2-35).

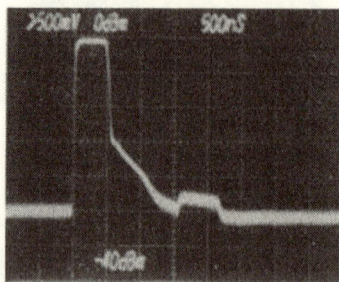

(a) $P_A = 0$ dBm $P_B = -40$ dBm.

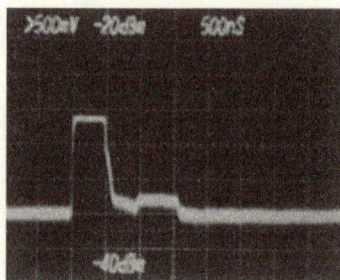

(b) $P_A = -20$ dBm $P_B = -40$ dBm.

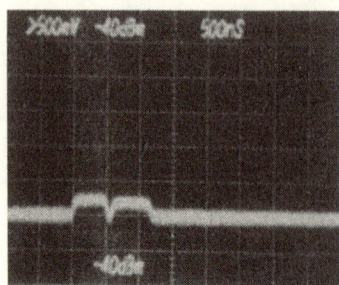

(c) $P_A = P_B = -40$ dBm.

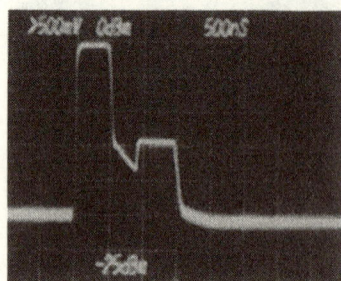

(d) $P_A = 0$ dBm $P_B = -25$ dBm.

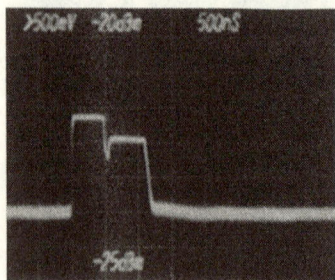

(e) $P_A = -20$ dBm $P_B = -25$ dBm.

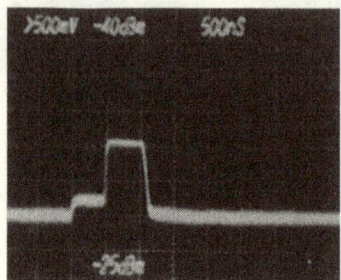

(f) $P_A = -40$ dBm $P_B = -25$ dBm.

FIGURE 5-5. Detector Logarithmic Video Recovery Time for Second Pulse Processing (see Figure 2-35).

98

If a thresholding comparator is to follow a logarithmic amplifier, the recovery time is that time necessary for the output signal to decrease to the signal threshold level. Figure 5-6 illustrates this definition. The solid constant line represents a threshold voltage. The threshold comparator will trigger when the first pulse exceeds this level and will trigger off when the signal drops below this threshold level, allowing the second signal to trigger the threshold comparator.

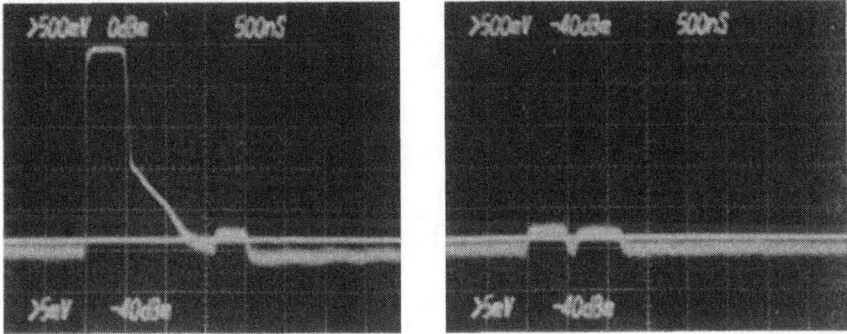

(a) P_A = 0 dBm P_B = -40 dBm. (b) P_A = P_B = -40 dBm.

FIGURE 5-6. Detector-Logarithmic Video Recovery Time for Signal Thresholding (see Figure 2-35).

A reasonable reference for recovery time measurement is 10% of the fall time. This is usually well-defined and easy to measure, as shown in Figure 5-7. Also shown is a reasonable definition of pulse width, i.e., pulse width $\cong$ 10% rise time to 10% fall time.

Maximum Duty Cycle

The classical definition of duty cycle is pulse width/PRF, and the maximum duty cycle for nonsaturated linear receivers is closely approximated by 1/pulse width. Logarithmic amplifiers have an intensity-dependent recovery time, and this somewhat complicates the definition of the maximum duty cycle for logarithmic receivers. Logarithmic IF amplifiers and DC–coupled detector–logarithmic amplifiers have a finite intensity–dependent recovery time (Figure 5–7), and the maximum duty cycle is given as

$$\text{Maximum duty cycle} = \frac{\text{Pulse width}}{\text{Pulse width} + \text{recovery time}} \qquad (5\text{–}3)$$

Log IF amplifiers have, in general, an advantage over detector–logarithmic video amplifiers.

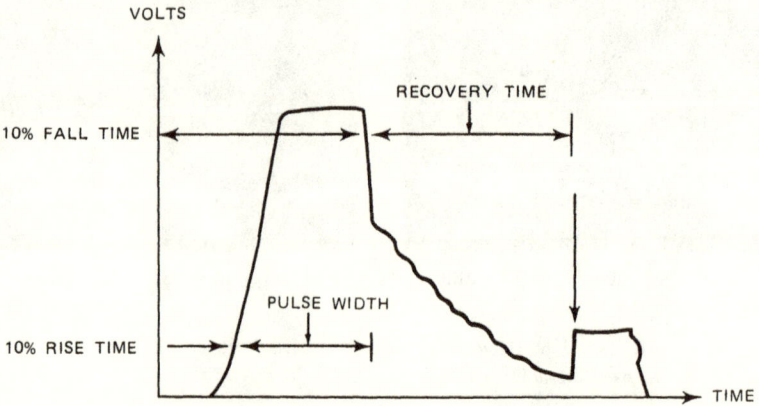

FIGURE 5–7. Definition of Recovery Time and Pulse Width.

Continuous Wave Rejection

The signal output of logarithmic amplifiers is susceptible to the presence of CW. The detector–logarithmic video amplifier can be designed to be insensitive to CW inputs (to the detector) of some –17 dBm or less (see Appendix H). The CW rejection for logarithmic IF amplifiers is directly dependent upon the received signal (see Appendix H) and may be approximated as

CW rejection (logarithmic IF) $\cong$ signal input (dBm) –3 $\qquad (5\text{–}4)$

Therefore, if the input signal is –35 dBm, the maximum CW input is –38 dBm.

100

Power Supply Rejection

The logarithmic transfer function and any DC offset errors for either logarithmic IF or detector–logarithmic video amplifiers may well be dependent upon the absolute DC power supply voltages. Some logarithmic amplifier suppliers include internal voltage regulators to minimize any external supply variations. In any case, the user and designer must account for any external DC supply variations dependent upon individual system requirements.

Warm-Up Time

The logarithmic conformity (and unit to unit matching) may depend on the time necessary for the individual transistors to achieve their operating temperature. This is design dependent and may range from milliseconds to minutes.

Chapter 6

SELECTED APPLICATIONS

Chapters 1 through 5 presented the basics of logarithmic amplification and characterization. This chapter presents several applications that have been published during the past several years. Because security awareness prevents discussion of recent configurations, those presented give a reasonable summary of logarithmic amplifiers currently used in the EW and radar world. The last two sections present the basics of extending the instantaneous input dynamic range of detector–logarithmic video amplifiers and a simple series lin–limit logarithmic IF technique.

Depending upon the application, the system engineer has a choice of using the logarithmic IF or the detector–logarithmic video amplifier. The previous chapters stressed what the author feels are the strong and weak points of each logarithmic technique. This chapter begins with a final comparison of logarithmic IF and detector–logarithmic video amplifiers.

Logarithmic IF/Detector–Logarithmic Video Comparison

Table 6-1 is a summary of logarithmic IF and detector–logarithmic video amplifiers. It should be noted that the weak points of each technique may well be improved in the future. Also, the data given are nominal and must not be treated as absolute.

TABLE 6-1. Logarithmic IF, Detector-Logarithmic Video
Amplifier Comparison.

Amplifier characteristics	Logarithmic IF	Detector logarithmic video
Maximum frequency, GHz	2	20
Bandwidth	500 MHz	16 GHz
Instantaneous input dynamic range, dBm	–75 to +10 (decreases as bandwidth increases)	Tunnel: –40 to +5 Schottky: –45 to +15
Log conformity, dB	±1	±0.5
Flatness, 500–MHz bandwidth, dB	±1.5	±0.5
Duty cycle, DC coupling	Easy	Moderate to difficult
CW rejection	Difficult	Easy
Pulse characteristics[a]	Excellent	Moderate
Unit-to-unit matching, dB	±1.5	±0.5
Cost	High (increases as bandwidth increases)	Moderate

[a]Minimum 1–dB pulsewidth and pulse recovery time (see Chapter 5).

Amplitude-Independent Frequency Discriminator

Figure 6-1 illustrates a simple amplitude-independent frequency discriminator. The filter outputs are logarithmically amplified and the voltages subtracted. The result is a linear frequency-dependent voltage (voltage is a linear function of logarithmic frequency) that is amplitude-insensitive.

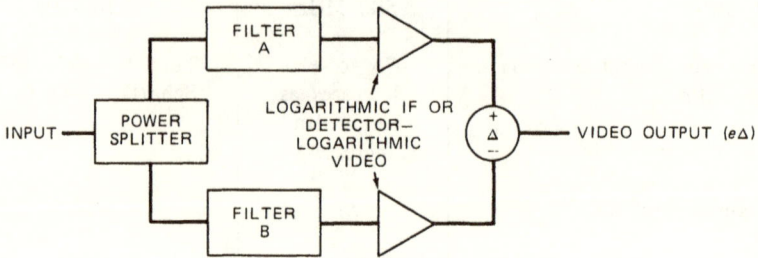

(a) Block diagram.

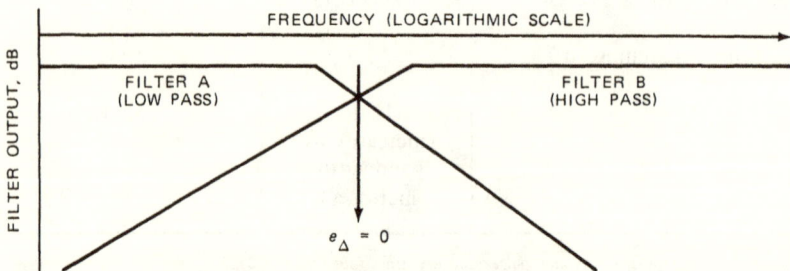

(b) Filter response.

FIGURE 6-1. Basic Frequency Discriminator.

Figure 6-2a illustrates another approach to an intensity-independent frequency discriminator that has been used in instantaneous frequency monitors [1]. This delay-line discriminator is described in detail by Wilkins and Kincheloe [2]. Figure 6-2b illustrates the discriminator characteristics.

104

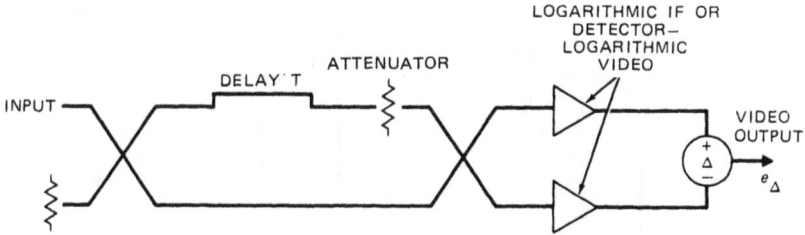

(a) Basic delay-line discriminator.

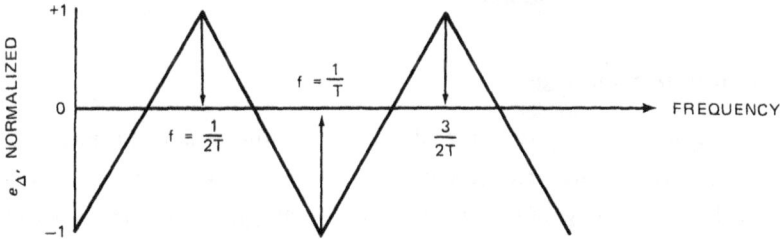

(b) Discriminator characteristics (attenuation = 0.5).

FIGURE 6-2. Basic Delay Line Discriminator.

Linearizing Automatic Gain Control Response

The AGC response (AGC time constant) is dependent upon the nonlinearities of the IF detector [3]. It was shown in Chapter 1 that a logarithmic video amplifier following a detector gives a linear voltage versus a dBm input response. This is used to advantage in the AGC circuit illustrated in Figure 6-3. A complete listing of AGC references is given in the bibliography.

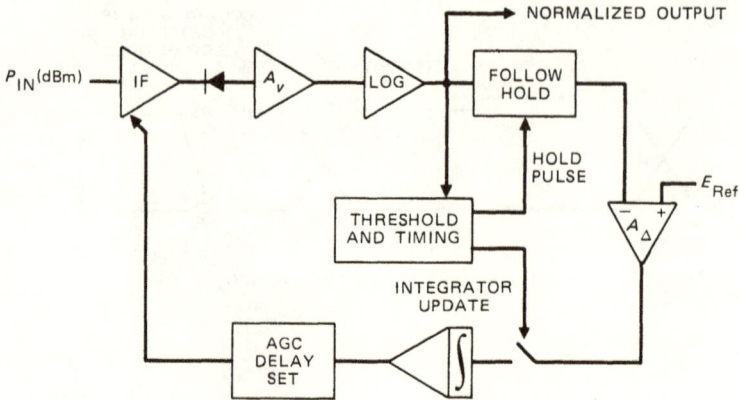

FIGURE 6-3. Linear AGC Response Loop.

Monopulse Direction Finding

The basic monopulse direction-finding concept discussed in Chapter 1 is one of the basic uses for logarithmic amplification. Figure 6-4 illustrates a two-channel monopulse DF [4 through 6] using the unique properties of the dual-mode spiral antenna [7]. Refer to the listed references for a detailed discussion of the monopulse technique.

The RF input is converted to an IF signal via the mixers and local oscillator. The IF signal is amplified via the variable gain IF amplifiers (to keep the signal level into the detectors in the middle of the logarithmic dynamic range). The signal is then converted into up-down and right-left direction-finding signals via the beam former [8]. The direction-finding signals are logarithmically amplified and held. The two difference signals represent the elevation and azimuth position of the received signal (see Chapter 1). The sum signal drives its own detector-logarithmic video for signal threshold detection and AGC normalization. The signal threshold comparator starts the necessary timing for the signal hold pulse and signal inhibit if a pseudo-DC-coupled detector-logarithmic video amplifier is used (see Figure 4-12), or if CW rejection is wanted (see Appendix H). Logarithmic IF amplifiers could replace the detector-logarithmic video amplifiers, provided the IF is low enough, to eliminate the need for AGC.

106

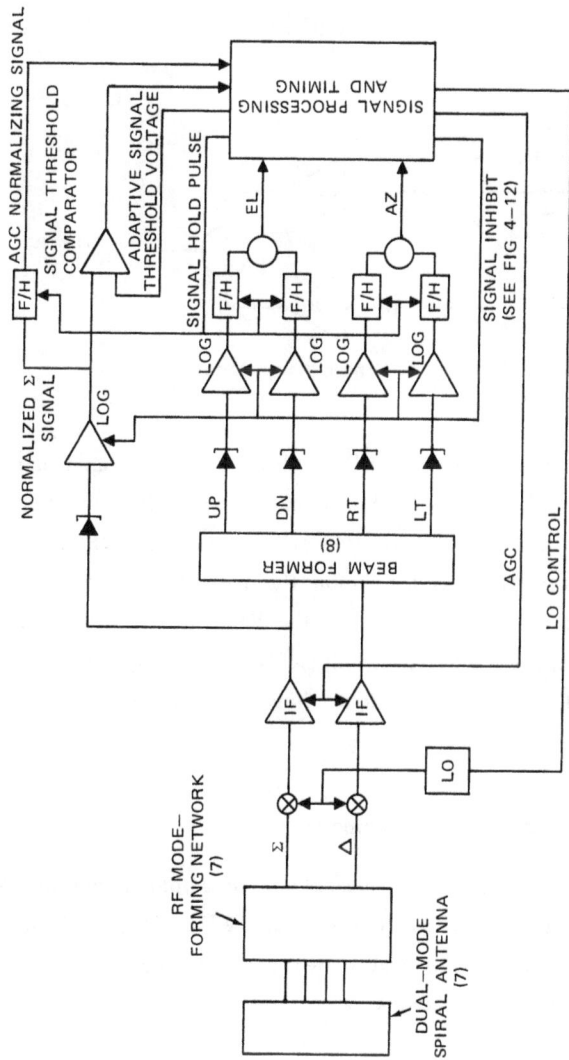

FIGURE 6-4. Basic Two-Channel Monopulse Direction-Finding Block Diagram.

NOTE: F/H = FOLLOW—HOLD AMPLIFIER

107

Extended Range Detector-Logarithmic Video Amplifier

The instantaneous input dynamic range of detector-logarithmic video amplifiers can be increased (at the expense of bandwidth, logarithmic conformity, and dollars) by using two detector-logarithmic videos offset by an RF amplifier as shown in Figure 6-5 [9].

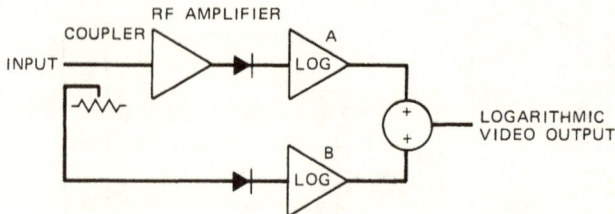

FIGURE 6-5. Basic Extended Range Detector-Logarithmic Video Amplifier.

The signal is amplified via the RF amplifier and is detected and logarithmically amplified by Log A. If the coupler loss is equal to the RF amplifier gain, detector-logarithmic Video B starts its logarithmic action when the RF amplifier limits. The logarithmic conformity is dependent upon the passband ripple of the RF amplifier and the coupler, on the limiting ripple of the RF amplifier, and on the temperature-dependent gain change of the RF amplifier. Dynamic ranges from –65 to +10 dBm have been achieved with this technique; however, the logarithmic conformity over temperature is typically ±2 dB.

A Simple "True" Logarithmic IF Technique

Chapter 3 discussed the "true" logarithmic IF amplifier concept using the series lin-limit logarithmic video amplifier concept discussed in Chapter 2 (see Figure 2-38). The simple circuit illustrated in Figure 6-6 performs the same function. The low-level amplifier gain is approximated as

$$A_V \cong \frac{-r_c}{r_e' + R_e} \tag{6-1}$$

where

r_c = effective AC collector resistance

r_e' = dynamic emitter resistance

and

$$r_e' = \frac{86.25 \times 10^{-6}\ T}{I_E} \quad \text{(see Equation 3-1)} \tag{6-2}$$

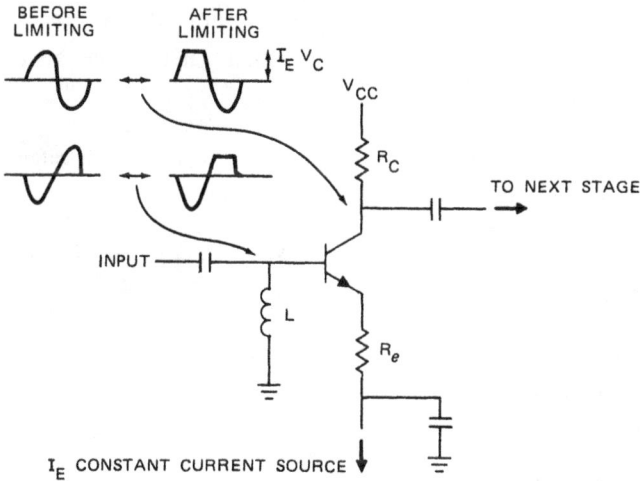

FIGURE 6-6. Basic "True" Logarithmic IF Stage.

The maximum positive output, e_L, is given as

$$e_L \cong I_E\ r_c \tag{6-3}$$

Thus, if the low-level gain is set at 2, the output voltage increses twice as fast as the input voltage (no surprise here) until the output limiting level, e_L, is reached. The gain is now unity (the positive going collector output voltage is limited), the output increases at the same rate as the input, and cascading such stages gives the wanted logarithmic response. The base inductor DC restores the voltage from the preceding stage. Filtering may be necessary to minimize the distortion caused by limiting. It is difficult to bias these logarithmic stages for optimum bandwidth giving a maximum IF of around 60 MHz or so.

109

References

1. Stanford University, Stanford Electronics Laboratories. *An Instantaneous Frequency-Measurement Receiver Specifically Developed for Automatic Signal Processing*, by J. M. Hunter. Stanford, Calif., October 1969. (Report No. 1969–4.)

2. –––––. *Microwave Realization of Broadband Phase and Frequency Discriminators*, by M. W. Wilkins and W. R. Kincheloe, Jr. Stanford, Calif., November 1968. (Report No. 1962/1966–2.)

3. Naval Weapons Center. *Automatic Gain Control: A Practical Approach to Its Analysis and Design*, by R. S. Hughes. China Lake, Calif., NWC, August 1977. (NWC TP 5948.)

4. McLendon, R. and C. Turner, "Broadband Sensors for Lethal Defense Suppression," *Microwave Journal*, September 1983, pp. 85-102.

5. Lipsky, S., "Find the Emitter Fast With Monopulse Methods," *Microwaves*, May 1978, pp. 42-53.

6. Bullock, L. G., G. R. OEH, and J. J. Sparagna, "An Analysis of Wide-Band Microwave Monopulse Direction Finding Techniques," *IEEE Transactions on Aerospace and Electronic Systems* Vol. A & D-7, No. 1 January 1971, pp. 188-202.

7. Mosko, J. A., "An Introduction to Wideband, Two-Channel Direction-Finding Systems," *Microwave Journal*, Part 1, February 1984, pp. 91-106 and Part II, March 1984, pp. 105-122.

8. U.S. Naval Ordnance Test Station. *Reduced Size, Dual-Mode Spiral for Two-Plane Monopulse Direction Finding*, by J. A. Mosko. China Lake, Calif., NOTS, May 1966. (NAVWEPS Report 8757, NOTS TP 3834.)

9. Sheade, M., "DLVAs Find Applications in ESM Systems," *Microwave System News*, August 1984 pp. 47-56.

Appendix A

BASIC DIFFERENTIAL AMPLIFIER THEORY

The differential amplifier illustrated in Figure A-1 is one of the most useful of all electronic circuits. It is the basic building block for many linear integrated circuits and in many of the logarithmic and linear video amplifier techniques presented.

This appendix presents the theory of the basic differential amplifier of Figure A-1.

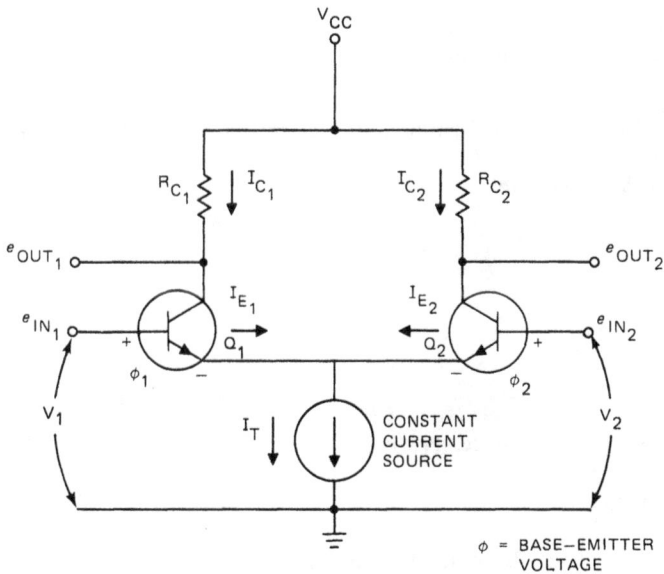

FIGURE A-1. Basic Differential Amplifier.

The DC and AC analyses for the basic differential amplifier will be found in terms of the Ebbers and Moll equations. It will be assumed that high β transistors are used. With reference to Figure A-1

$$I_{E_1} = I_S \left[\exp \left(\frac{\phi_1}{V_T} \right) - 1 \right] \tag{A-1a}$$

$$I_{E_2} = I_S \left[\exp \left(\frac{\phi_2}{V_T} \right) - 1 \right] \tag{A-1b}$$

where

I_S = transistor saturation current

$$V_T = \frac{KT}{q}$$

K = Bolzmann's constant
q = electron charge
T = temperature in degrees K
ϕ = base–emitter voltage drop

If $\exp \dfrac{\phi}{V_T} \gg 1$ (this is legitimate for ϕ greater than several millivolts), Equations A-1a and A-1b become

$$I_{E_1} = I_S \exp \left(\frac{\phi_1}{V_T} \right) \tag{A-2a}$$

$$I_{E_2} = I_S \exp \left(\frac{\phi_2}{V_T} \right) \tag{A-2b}$$

The constant current source, I_T, is given as

$$I_T = I_{E_1} + I_{E_2} \tag{A-3}$$

Summing the voltage drops in the base-emitters

$$V_1 + \left(\pm e_{in_1} \right) - \phi_1 + \phi_2 - V_2 - \left(\pm e_{in_2} \right) = 0 \tag{A-4}$$

112

where

$\pm e_{in}$ = input pulse amplitudes (+ for positive and – for negative)

The DC analysis is performed first, thus $e_{in} = 0$

Solving Equation A-4 for ϕ_2

$$\phi_2 = V_2 + \phi_1 - V_1 \tag{A-5}$$

Substituting Equations A-5 and A-2b into Equation A-3

$$I_T = I_{E_1} + I_S \exp\left[\frac{(V_2 + \phi_1 - V_1)}{V_T}\right] \tag{A-6a}$$

and using Equation A-2a

$$I_T = I_{E_1} + I_{E_1} \exp\left[\frac{(V_2 - V_1)}{V_T}\right] \tag{A-6b}$$

Solving for I_{E_1}

$$I_{E_1} = \frac{I_T}{1 + \exp\left[\frac{(V_2 - V_1)}{V_T}\right]} \tag{A-7}$$

The collector current is given as (since $\beta \gg 1$)

$$I_{C_1} = I_{E_1} \tag{A-8a}$$

Thus

$$I_{C_1} = \frac{I_T}{1 + \exp\left[\frac{(V_2 - V_1)}{V_T}\right]} \tag{A-8b}$$

Using the same procedure

$$I_{E_2} = \frac{I_T}{1 + \exp\left[\dfrac{(V_1 - V_2)}{V_T}\right]} \qquad\qquad\text{(A-9)}$$

or, again since $a \cong 1$

$$I_{C_2} = \frac{I_T}{1 + \exp\left[\dfrac{(V_1 - V_2)}{V_T}\right]} \qquad\qquad\text{(A-10)}$$

Figure A-2 illustrates Equations A-7 and A-9 ($I_T = 1$) for three temperatures. Several facts should be noticed from Figure A-2.

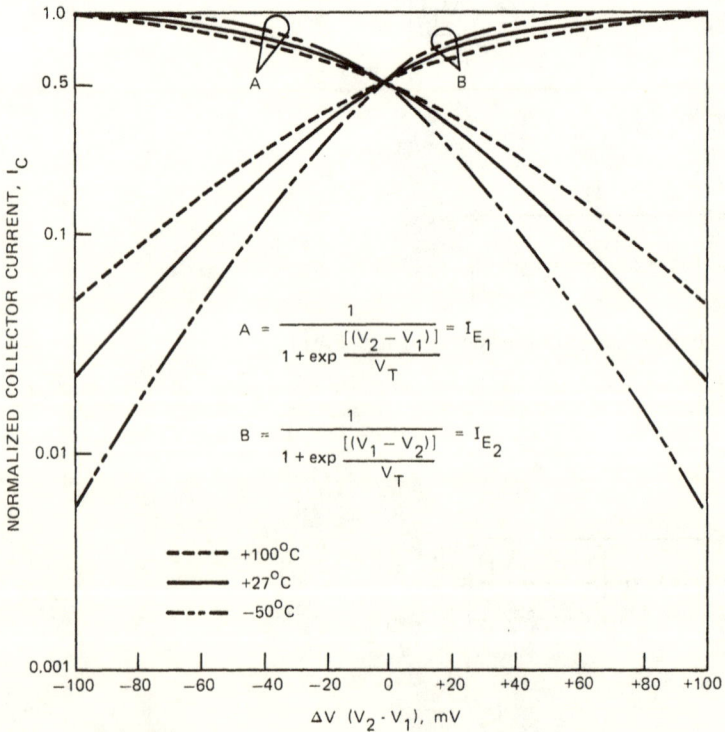

FIGURE A-2. Differential Amplifier DC Characteristics.

1. For a given $\left|\Delta V\right|$, $I_{E_1} = I_{E_2}$.

2. The temperature coefficient for $\Delta V = 0$ is 0.

3. The current temperature coefficient for $I_E > 0.5$ is negative (as T increases I_E decreases), and the current temperature coefficient for $I_E < 0.5$ is positive (increasing T increasing I_E).

4. The voltage temperature coefficient for $I_E > 0.5$ is positive (it is also positive for $I_E < 0.5$).

5. The current reaches a maximum for $\Delta V \cong 100$ mV.

6. $I_{E_1} + I_{E_2} = I_T$ (from Equation A-3).

The AC analysis will be performed first in terms of e_{in_1} and then for e_{in_2}.

Substituting Equations A-4 and A-2b into Equation A-6a (letting $e_{in_2} = 0$)

$$I_T = I_{C_1} + I_s \exp \left\{ \left[\frac{V_2 + \phi_1 - V_1 - (\pm e_{in})}{V_T} \right] \right\} \tag{A-11}$$

using Equation A-2a and solving for I_{C_1}

$$I_{C_1} = \frac{I_T}{1 + \exp \left[\left(\frac{V_2 - V_1 - \left[\pm e_{in} \right]}{V_T} \right) \right]} \tag{A-12}$$

Because of the bias voltage, the I_{C_1} current is (from Equation A-8b)

$$I_{C_1} \Big|_B = \frac{I_T}{1 + \exp \left[\frac{(V_2 - V_1)}{V_T} \right]} \tag{A-13}$$

Let $V_2 - V_1 = \Delta V$ \hfill (A-14)

and Equation A-13 becomes

$$I_{C_1} \Big|_B = \frac{I_T}{1 + \exp \frac{(\Delta V)}{V_T}} \tag{A-15}$$

115

The output pulse is given as

$$e_{out_1} = E_{out_{total}} - E_{out_{bias}} \tag{A-16}$$

where

$$E_{out_{total}} = V_{CC} - I_{C_1} R_{C_1} \tag{A-17}$$

$$E_{out_{bias}} = V_{CC} - \left(I_{C_1} \Big|_B \right) R_{C_1} \tag{A-18}$$

Using Equations A-12, A-15, A-16, A-17, and A-18

$$e_{out_1} = V_{CC} - \cfrac{I_T R_{C_1}}{1 + \left[\exp \cfrac{(\Delta V)}{V_T} \right] \exp \left[- \cfrac{(\pm e_{in})}{V_T} \right]}$$

$$- \left[V_{CC} - \cfrac{I_T R_{C_1}}{1 + \exp \cfrac{(\Delta V)}{V_T}} \right] \tag{A-19}$$

Let

$$\exp \frac{(\Delta V)}{V_T} = K_a \tag{A-20}$$

Equation A-19 now becomes

$$e_{out_1} = I_T R_{C_1} \left(\cfrac{K_a \left\{ \exp \left[- \cfrac{(\pm e_{in})}{V_T} \right] - 1 \right\}}{(K_a + 1) + K_a (K_a + 1) \exp \left[- \cfrac{(\pm e_{in})}{V_T} \right]} \right) \tag{A-21}$$

The outputs for the individual input polarities may be given as (for $e_{in_2} = 0$)

Positive input:

$$e_{out_1}\Big|_+ = -\, I_T R C_1 \left\{ \frac{K_a \left[\exp \dfrac{(e_{in})}{V_T} - 1 \right]}{K_a (K_a + 1) + (K_a + 1) \exp \dfrac{(e_{in})}{V_T}} \right\} \qquad (A-22)$$

Negative input:

$$e_{out_1}\Big|_- = I_T R C_1 \left\{ \frac{K_a \left[\exp \dfrac{(e_{in})}{V_T} - 1 \right]}{(K_a + 1) + K_a (K_a + 1) \exp \dfrac{(e_{in})}{V_T}} \right\} \qquad (A-23)$$

The same analysis may be performed to determine the output e_{out_2} yielding

Positive input:

$$e_{out_2}\Big|_+ = I_T R C_2 \left\{ \frac{K_b \left[\exp \dfrac{(e_{in})}{V_T} - 1 \right]}{(K_b + 1) + K_b (K_b + 1) \exp \dfrac{(e_{in})}{V_T}} \right\} \qquad (A-24)$$

Negative input:

$$e_{out_2}\Big|_- = -\, I_T R C_2 \left\{ \frac{K_b \left[\exp \dfrac{(e_{in})}{V_T} - 1 \right]}{K_b (K_b + 1) + (K_b + 1) \exp \dfrac{(e_{in})}{V_T}} \right\} \qquad (A-25)$$

where

$$K_b = \exp \left[\frac{(V_1 - V_2)}{V_T} \right] = \frac{1}{K_a} \qquad (A-26)$$

117

If $R_{C_1} = R_{C_2}$

$$e_{out_1}\Big|_+ = - e_{out_2}\Big|_+ \tag{A-27}$$

and

$$e_{out_1}\Big|_- = - e_{out_2}\Big|_- \tag{A-28}$$

The outputs for e_{in_2} ($e_{in_1} = 0$) may be found in a similar fashion. The equations are identical to Equations A-22 and A-23, with the negative inputs replaced by positive.

Letting $R_{C_1} = R_{C_2}$, and substituting $K_b = \dfrac{1}{K_a}$

$$e_{out_1}\Big|_+^{e_{in_1}} = - e_{out_2}\Big|_+^{e_{in_1}} = e_{out_1}\Big|_-^{e_{in_2}} = - e_{out_2}\Big|_-^{e_{in_2}}$$

$$= I_T R_C \left\{ \frac{K_a \left[\exp \dfrac{(e_{in})}{V_T} - 1 \right]}{K_a (K_a + 1) + (K_a + 1) \exp \dfrac{(e_{in})}{V_T}} \right\} \tag{A-29}$$

where

$$e_{out_1}\Big|_+^{e_{in_1}} = \text{output for a positive input on } e_{in_1}, \text{ etc.} \tag{A-30}$$

$$e_{out_1}\Big|_-^{e_{in_1}} = - e_{out_2}\Big|_-^{e_{in_1}} = e_{out_1}\Big|_+^{e_{in_2}} = - e_{out_2}\Big|_+^{e_{in_2}}$$

$$= I_T R_C \left\{ \frac{K_a \left[\exp \dfrac{(e_{in})}{V_T} - 1 \right]}{(K_a + 1) + K_a (K_a + 1) \exp \dfrac{(e_{in})}{V_T}} \right\} \tag{A-31}$$

118

The maximum outputs $\left(e_{\text{out}_1} \text{ or } e_{\text{out}_2}\right)$ may be given as

$$e_{\text{out}_1}\bigg|_{+\text{max}}^{e_{\text{in}_1}} = - e_{\text{out}_2}\bigg|_{+\text{max}}^{e_{\text{in}_1}} = e_{\text{out}_1}\bigg|_{-\text{max}}^{e_{\text{in}_2}} = - e_{\text{out}_2}\bigg|_{-\text{max}}^{e_{\text{in}_2}}$$

$$= \underset{e_{\text{in}} \to \infty}{\text{Lim}}\ e_{\text{out}} = - \frac{I_T R_C K_a}{K_a + 1} \tag{A-32}$$

$$e_{\text{out}_1}\bigg|_{-\text{max}}^{e_{\text{in}_1}} = - e_{\text{out}_2}\bigg|_{-\text{max}}^{e_{\text{in}_1}} = e_{\text{out}_1}\bigg|_{+\text{max}}^{e_{\text{in}_2}} = - e_{\text{out}_2}\bigg|_{+\text{max}}^{e_{\text{in}_2}}$$

$$= \underset{e_{\text{in}} \to \infty}{\text{Lim}}\ e_{\text{out}} = - \frac{I_T R_C}{K_a + 1} \tag{A-33}$$

Figure A-3 illustrates the effect of ΔV ($K_a = \exp(\Delta V / V_T)$) on Equation A-29 and A-31.

Most designs require that the offset voltage, ΔV, be zero. Thus, $K_a = K_b = 1$, and Equations A-29 and A-31 become:

$$\left| e_{\text{out}_1} \right| = \frac{I_T R_{C_1}}{2} \left[\frac{\exp(e_{\text{in}}/V_T) - 1}{\exp(e_{\text{in}}/V_T) + 1} \right] \tag{A-34}$$

$$\left| e_{\text{out}_2} \right| = \frac{I_T R_{C_2}}{2} \left[\frac{\exp(e_{\text{in}}/V_T) - 1}{\exp(e_{\text{in}}/V_T) + 1} \right] \tag{A-35}$$

If $R_{C_1} = R_{C_2}$, Equations A-34 and A-35 reduce to

$$\left| e_{\text{out}} \right| = \frac{I_T R_C}{2} \left[\tanh\left(\frac{e_{\text{in}}}{2 V_T} \right) \right] \tag{A-36}$$

119

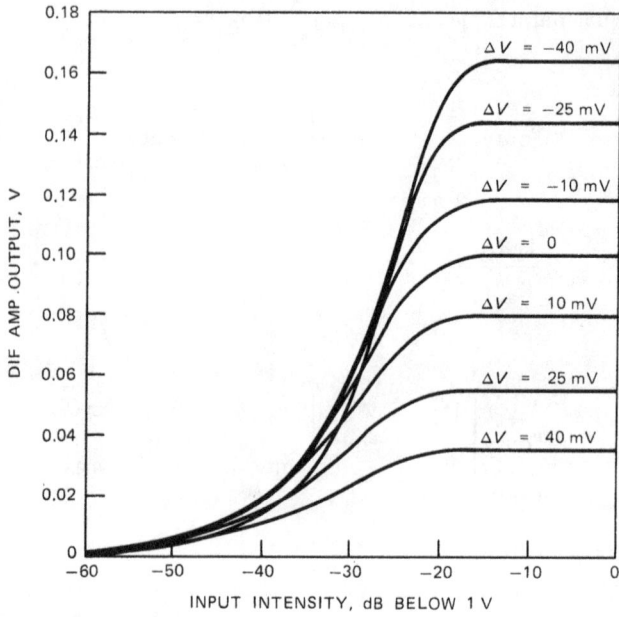

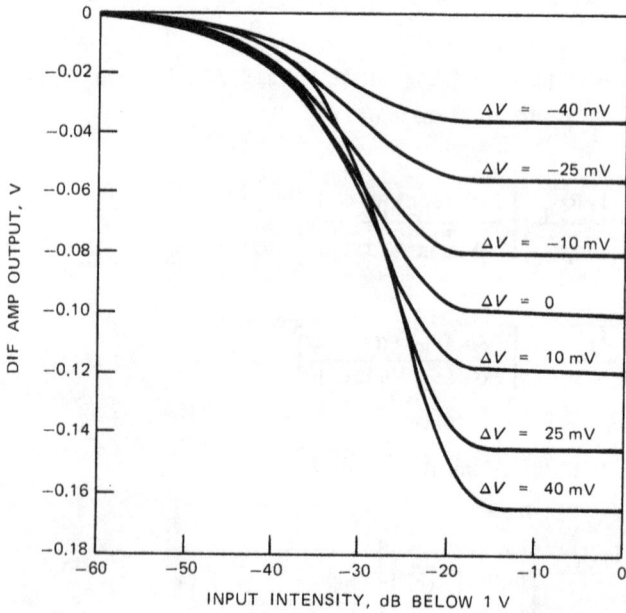

FIGURE A-3. Differential Amplifier Output for $I_T R_C = 0.2$.

If $e_{in}/2V_T$ is small

$$\left| e_{out} \right| = \frac{I_T R_C}{4V_T} \, e_{in}$$

or the low level voltage gain is given as

$$A_V = \frac{I_T R_C}{4V_T}$$

Appendix B

ON VARIOUS FACTORS AFFECTING THE LOGARITHMIC ACCURACY (LOGARITHMIC CONFORMITY) AND MATCHING OF PARALLEL SUMMATION LOGARITHMIC VIDEO AMPLIFIERS USING A DIFFERENTIAL AMPLIFIER AS THE LOGARITHMIC ELEMENT

Most (if not all) present-day logarithmic video amplifiers are really approximations to a true logarithmic response, employing the differential amplifier as the pseudologarithmic element. Figure B-1 illustrates the basic configuration. Differential amplifier pseudologarithmic elements have a fairly well-defined logarithmic range, as shown in Figure B-2. Offsetting the individual logarithmic elements by the proper gains (see Figure B-1) results in a well-defined logarithmic response, as shown in Figure B-3. The theoretical logarithmic error is a function of the logarithmic range each logarithmic element supplies (dB/stage). This logarithmic error for various differential amplifier logarithmic-element dynamic ranges is given in Table B-1. It is given in ±dB (voltage input) and ±dBm when used with square law detectors. Thus, for a 15–dB/stage, each logarithmic element in Figure B-1 must be displaced by 15 dB of gain (or attenuation).

TABLE B-1. Logarithmic Error for Various Differential
Amplifier Logarithmic-Element Dynamic Ranges.

dB/Stage	Logarithmic error	
	±dB	±dBm (square law detector)
20	0.8	0.4
15	0.4	0.2
10	0.2	0.1

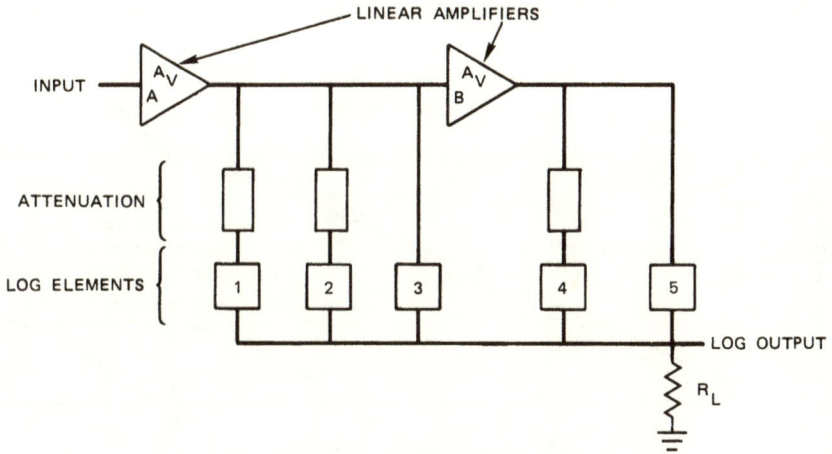

FIGURE B-1. Basic Logarithmic Video Configuration.

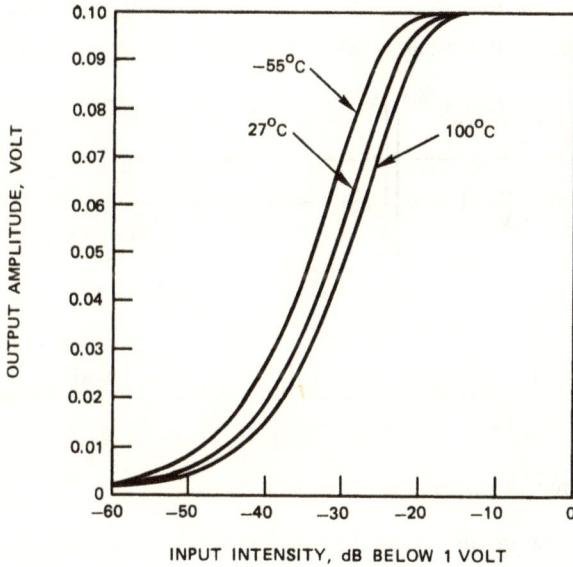

FIGURE B-2. Output Versus Input for a Differential
Amplifier Logarithmic Element, Where $I_T R_L/2 = 0.1$.

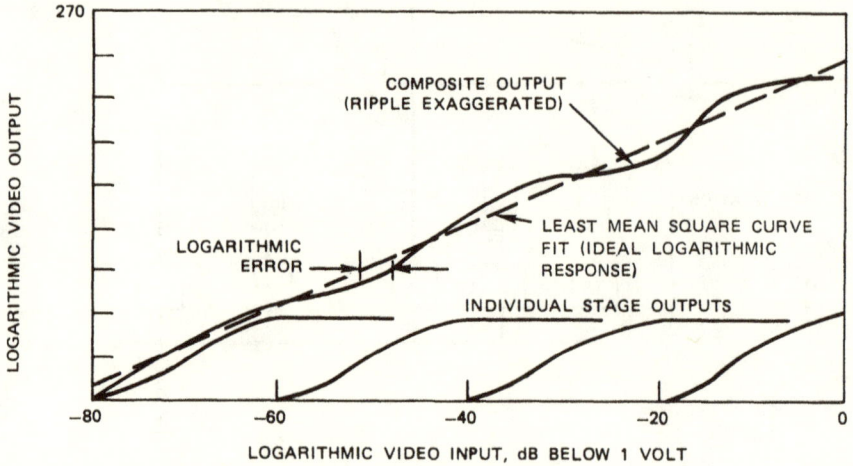

FIGURE B-3. Composite Output of a Four-Stage Pseudologarithmic Amplifier.

As mentioned, the logarithmic element most often used is the differential amplifier illustrated in Figure B-4. The pulse output voltage for this circuit may be given, after simplifying Equations A–24 and A–30, as

$$e_O = I_T R_L \left[\frac{\exp\left(\frac{\Delta V}{V_T}\right)}{\exp\left(\frac{\Delta V}{V_T}\right) + 1} \right] \left[\frac{\exp\left(\frac{e_{LE}}{V_T}\right) - 1}{\exp\left(\frac{e_{LE}}{V_T}\right) + \exp\left(\frac{\Delta V}{V_T}\right)} \right] \tag{B-1}$$

where

$\Delta V = V_{B2} - V_{B1}$ [or the difference in DC base voltages]

$V_T = \dfrac{KT}{q}$ [or $86.25 \times 10^{-6}(273 + °C)$] $\cong 0.026$ [at $27°C$]

e_{LE} = logarithmic element input voltage

The imposing nature of Equation B-1 reduces to a well-defined relationship if $\Delta V = 0$.

$$\left. e_O \right|_{\Delta V = 0} = \frac{I_T R_L}{2} \text{Tanh} \frac{e_{LE}}{2V_T} \tag{B-2}$$

124

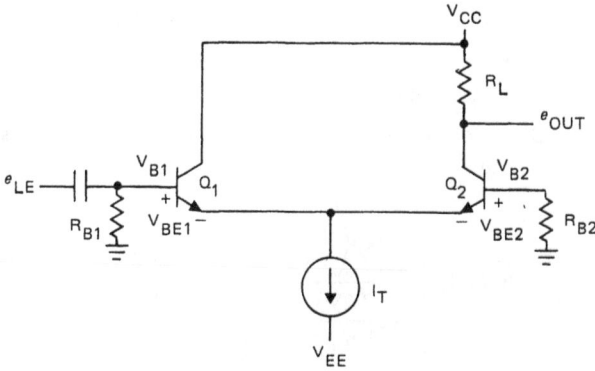

FIGURE B-4. Basic Differential Amplifier Logarithmic Element.

Figure B-5 illustrates Equation B-1 for various values of ΔV. If $\Delta V = 0$, certainly an ideal approximation that will be discussed shortly, the output has a well-defined limiting value of

$$e_O \bigg|_{\substack{\text{Limit} \\ \Delta V = 0}} = \pm \frac{I_T R_L}{2} \tag{B-3}$$

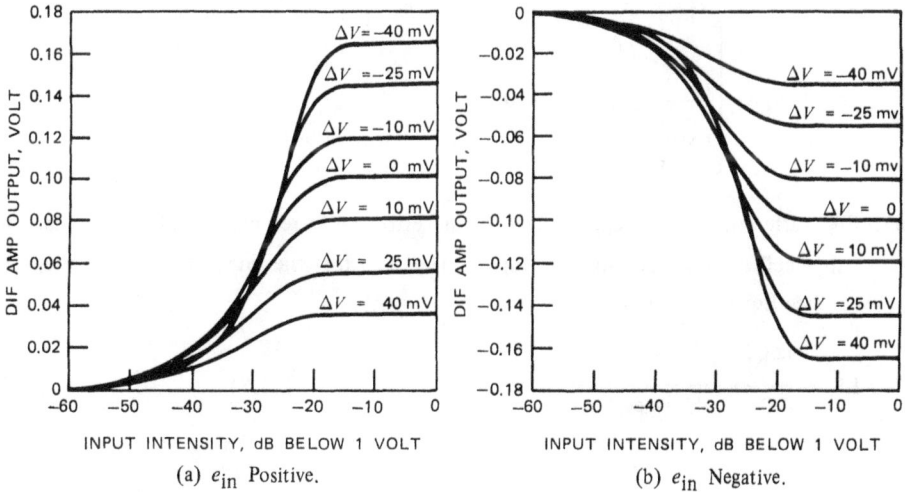

(a) e_{in} Positive.

(b) e_{in} Negative.

FIGURE B-5. Differential Amplifier Logarithmic-Element Characteristics

(Equation B-1) $\dfrac{I_T R_L}{2} = 0.1$

125

As can be seen from Figure B-5, for ΔV greater than 10 mV, the deviation in the output-limiting level is excessive (and the logarithmic error is a direct function on this limiting level). Assuming a ΔV of less than 5 mV, the second bracketed term in Equation B-1 closely approximates a Tanh function. Thus, Equation B-1 may be written as

$$e_0 \Big|_{\Delta V < 5 \text{ mV}} \cong I_T R_L \left[\frac{\exp\left(\dfrac{\Delta V}{V_T}\right)}{\exp\left(\dfrac{\Delta V}{V_T}\right) + 1} \right] \text{Tanh } \frac{e_{LE}}{2V_T} \tag{B-4}$$

and the limiting output becomes

$$e_L = I_T R_L \left[\frac{\exp\left(\dfrac{\Delta V}{V_T}\right)}{\exp\left(\dfrac{\Delta V}{V_T}\right) + 1} \right] \tag{B-5}$$

The difference in limiting level between Equation B-5 ($\Delta V \neq 0$) and the ideal of Equation B-3 ($\Delta V = 0$) is

$$\Delta e_L = I_T R_L \left[\frac{\exp\left(\dfrac{\Delta V}{V_T}\right)}{\exp\left(\dfrac{\Delta V}{V_T}\right) + 1} - 0.5 \right] \tag{B-6}$$

and this difference in voltage represents an error with respect to ideal that can be given in decibels referred to the input. The LS for the ideal case ($\Delta V = 0$) is given as (see Equation 2-16)

$$LS = \frac{I_T R_L}{2} \left(\frac{1}{dB/stage} \right) \frac{volts}{dB} \tag{B-7}$$

where

dB/stage = individual logarithmic-element dynamic range (see Table B-1)

$\dfrac{I_T R_L}{2}$ = limiting level for $\Delta V = 0$

Dividing Equation B-6 by Equation B-7 yields the decibel error, with respect to ideal, as a function of ΔV.

$$\text{error} = \frac{\Delta e_L}{LS} \; dB \tag{B-8}$$

or

$$\text{error} = 2 \left(\frac{dB}{\text{stage}} \right) \left(\frac{\exp\left(\dfrac{\Delta V}{V_T}\right)}{\exp\left(\dfrac{\Delta V}{V_T}\right) + 1} - 0.5 \;\; dB \right) \tag{B-9}$$

The error as a function of ΔV is given in Table B-2 (for $T = 27°C$).

TABLE B-2. Decibel Error as a Function of
ΔV for Several dB/Stages.

ΔV, ± mV	Logarithmic error, ±dB*		
	10-dB/stage	15-dB/stage	20-dB/stage
0	0.0	0.0	0.0
1	.19	.29	0.38
2	.38	.58	0.77
3	.58	0.86	1.15
4	.77	1.15	1.53
5	0.96	1.44	1.92

*When used with square law detector, divide the logarithmic error by 2 (see Appendix F).

To put Equation B-9 and Table B-2 in perspective, if one has two matched logarithmic amplifiers (each with a 15-dB/stage), and one logarithmic element has a 3-mV DC offset voltage at its base, the two amplifiers will have a 0.86-dB error between them. The obvious point is that for matched pairs, any DC voltages at the bases of the differential amplifier logarithmic elements must be quite small (or be equal). This is fairly simple if the logging elements are AC-coupled to their linear driving amplifiers (see Figure B-1). With reference to Figure B-4, the base voltages may be given as

$$V_{B1} = \frac{-I_T R_{B1}}{2\,(\beta_1 + 1)} \tag{B-10a}$$

$$V_{B2} = \frac{-I_T R_{B2}}{2\,(\beta_2 + 1)} \tag{B-10b}$$

If the two transistors are within an integrated circuit, the following tolerances can be expected

$$\Delta\beta < 10\%$$

$$\Delta R_B < 2\%$$

$$\beta|_{min} = 80 \qquad \beta|_{typ} = 180 \qquad \beta|_{max} = 300$$

Thus, for an $R_B = 100\ \Omega$ and $I_T = 1$ mA.

$$V_B\big|_{typ} = -0.276\ \text{mV} \quad (\beta = 180,\ R_B = 100\ \Omega) \tag{B-11}$$

$$V_B\big|_{max} = -0.630\ \text{mV} \quad (\beta = 80,\ R_B = 102\ \Omega) \tag{B-12}$$

$$V_B\big|_{min} = -0.163\ \text{mV}, \quad (\beta = 300,\ R_B = 98\ \Omega) \tag{B-13}$$

Thus, the worst case ΔV is 0.47 mV, which results in an error (Equation B-9 and Table B-2) of 0.14 dB (for a 15-dB/stage). Note that this ΔV is the worst case possible, a situation not likely to happen. Thus, little logarithmic error is to be expected for AC-coupled logarithmic elements if R_B is held to 100 Ω or so.

DC-coupling the logarithmic elements places extreme DC tolerance requirements on the linear video amplifiers (see Figure B-1), as only a 3-mV DC output from one of the linear stages will give, for a 15-dB/stage logarithmic element, an 0.85-dB error.*

The differential amplifier current source, I_T, also has a direct impact on the limiting level and thus the logarithmic error. Assuming $\Delta V = 0$, the difference in limiting level for different values of I_T (I_T and I_{TX}) is

$$e_L\Big|_{ideal} = \frac{I_T R_L}{2} \tag{B-14}$$

$$e_L\Big|_{I_{TX}} = \frac{I_{TX} R_L}{2} \tag{B-15}$$

Thus

$$\Delta e_L = \frac{R_L}{2}(I_{TX} - I_T) \tag{B-16}$$

The dB error due to Δe_L may now be found by dividing Equation 16 by the logarithmic slope (Equation B-7).

$$\text{Error} = \frac{\dfrac{R_L}{2}(I_{TX} - I_T)}{\dfrac{I_T R_L}{2}\left(\dfrac{1}{dB/stage}\right)} \ dB \tag{B-17}$$

or

$$\text{Error} = \frac{\Delta I_T}{100}\left(\frac{dB}{stage}\right) \ dB \tag{B-18}$$

where

$$\Delta I_T = \frac{I_{TX} - I_T}{I_T} \times 100$$

Table B-3 illustrates the error as a function of this current match.

*Again, when used with square law detectors, dBm = dB/2, or, for the case at hand, the error at the detector input is 0.425 dBm.

129

TABLE B-3. Logarithmic Error as a Function of $\Delta I_T\%$.

ΔI_T, %	Logarithmic error, dB		
	10-dB/stage	15-dB/stage	20-dB/stage
1	0.1	0.15	0.2
3	.3	0.45	0.6
5	.5	0.75	1.0
7	0.7	1.05	1.4
10	1	1.5	2

There are various techniques for generating I_T (see Appendix C); however, Equation B-18 (and Table B-3) should be useful in determining the best approach for a given application.

Referring again to Figures B-1, B-4, and Equation B-2, the input to any given logarithmic element (e_{LE}) is

$$e_{LE} = A_V\, e_{in} \qquad\qquad (B-19)$$

where

e_{in} = logarithmic amplifier input

A_V = gain (or attenuation) proceding the logarithmic element

Substituting Equation B-19 into Equation B-2,

$$e_o = \frac{I_T R_L}{2}\, \text{Tanh}\, \frac{A_V e_{in}}{2 V_T} \qquad\qquad (B-20)$$

Thus, to ensure a small logarithmic error between two logarithmic amplifiers, the voltage gain preceding each logarithmic element must be well-matched. The effect of A_V on Equation B-20 is simply to shift the curve with respect to e_{in} as illustrated in Figure B-3. (After all, this shift in the individual logarithmic element curve with A_V is what this concept is all about.) Thus, a shift of 1 dB in A_V will

130

give a 1-dB error. It should be noted that A_V does not affect the limiting level; thus, any error associated with the A_V preceding a given logarithmic element will only cause an error during the time the stage is active. (For a logarithmic element of a 15 dB/stage, the errors are only a problem during the 15 dB that the stage is active.) Fortunately, matching A_V to within a percent or so poses little difficulty.

The effect V_T, Equation B-20, has on the logarithmic error is the inverse of A_V. Any increase in V_T effectively decreases the voltage gain (A_V/V_T decreases as V_T increases), effectively shifting the individual logarithmic element curves. Increasing V_T by 1 dB has the same effect as decreasing A_V by 1 dB. V_T has been defined as

$$V_T = \frac{KT}{q} \qquad\qquad (B-21)$$

where

K = Boltzmann's constant
q = electron charge
T = temperature in degrees Kelvin

V_T may now be given as

$$V_T = 86.25 \times 10^{-6} \, (T) \qquad\qquad (B-22)$$

where

T = 273 + °C

The change in V_T with T is

$$\Delta V_T = 86.25 \times 10^{-6} \, \Delta T \qquad\qquad (B-23)$$

and this temperature difference, ΔT, simply shifts the individual logarithmic element curve. It should be noted that for a given theoretically perfect logarithmic amplifier using the differential amplifier as the logarithmic element, the LS will be insensitive to temperature; however, the curve will shift as temperature is changed. Table B-4 gives the shift in logarithmic response as a function of ΔT (in °C).

$\pm\Delta T/$°C*	$\pm$Logarithmic error, dB
1	0.03
3	0.087
5	0.146
10	0.294

*27°C (300°K) is used as the reference temperature.

Table B-4 also gives two most important results. First, if two logarithmic elements are within 3°C or so, little logarithmic error will be noticed. (If both logarithmic elements are on the same integrated circuit chip, this is an easy criterion to meet with careful layout procedures.) Second, for a given logarithmic amplifier, the logarithmic curve will shift slightly with temperature, and, if necessary, this shift can be compensated for by decreasing the gain of amplifier A in Figure B-1 as the temperature is increased [1]. (From Table B-4, decrease the gain 0.03 dB per °C.)

Another important source of logarithmic error is the differential amplifier transistors (see Figure B-4). The collector current for a transistor may be given as

$$I_C = I_S \exp \frac{V_{BE}}{V_T} \tag{B-24}$$

where I_S is a constant used to describe the transfer characteristic of the transistor (typical values for modern integrated transistors are on the order of 10^{-14} amps). The development of Equation B-1 assumed that I_S for both transistors was equal, which in actuality is not necessarily true. The effect of I_S on the logarithmic error will now be covered. With reference to Figure B-4, if the DC voltage drop across R_{B1} and R_{B2} is small (a reasonable assumption for small-valued resistors), the base-emitter voltages will be equal. The collector current of Q_1 may now be given as (assuming that both transistors are at the same temperature)

$$I_{C1} = I_{S1} \exp \frac{V_{BE}}{V_T} \tag{B-25}$$

and I_{C2} may be given as

$$I_{C2} = I_{S2} \exp \frac{V_{BE}}{V_T} \tag{B-26}$$

Ideally, $I_{S1} = I_{S2}$, and the two collector currents are equal, which is not necessarily the case. Taking the ratio of I_{C1} to I_{C2}

$$\frac{I_{C1}}{I_{C2}} = \frac{I_{S1}}{I_{S2}} \tag{B-27}$$

Since

$$I_T \cong I_{C1} + I_{C2}, \tag{B-28}$$

$$I_{C2} = I_T \left(\frac{1}{1 + \dfrac{I_{S1}}{I_{S2}}} \right) \tag{B-29}$$

The error, with respect to ideal (Equation B-3), in the limited output is now given as

$$\Delta e_L = \left[\frac{I_T}{2} - I_T \left(\frac{1}{1 + \dfrac{I_{S1}}{I_{S2}}} \right) \right] R_L \tag{B-30}$$

or

$$\Delta e_L = \frac{I_T R_L}{2} \left[\frac{\dfrac{I_{S1}}{I_{S2}} - 1}{\dfrac{I_{S1}}{I_{S2}} + 1} \right] \tag{B-31}$$

Dividing the limiting error of Equation B-31 by the LS (Equation B-7), the logarithmic error may be given as

$$\text{Error} = \left(\frac{dB}{\text{stage}} \right) \left(\frac{\dfrac{I_{S1}}{I_{S2}} - 1}{\dfrac{I_{S1}}{I_{S2}} + 1} \right) dB \tag{B-32}$$

Table B-5 lists the logarithmic error as a function of I_S match (in %), where

$$\Delta I_S \Big|^{\%} = \left(\frac{I_{S1}}{I_{S2}} - 1 \right) 100 \tag{B-33}$$

The match in I_S is directly dependent on the photolithographic resolution limits of the equipment used in the transistor layout. (I_S is directly dependent on the transistor emitter area; thus, the better the area resolution, the better the I_S match.)

TABLE B-5. Logarithmic Error as a
Function of ΔI_S (%).

$\Delta I_S \mid \%$	Logarithmic error, dB		
	10–dB/stage	15–dB/stage	20–dB/stage
1	0.05	0.07	0.1
5	0.2	0.3	0.4
10	0.5	0.7	1.0

References

1. Jaffe, S. "Temperature Compensation Using Thermistors," *Microwaves and RF,* April 1984, pp. 101–104.

Appendix C

CONSTANT CURRENT SOURCES

Logarithmic video amplifiers employing differential amplifiers as the logarithmic element, require temperature insensitive current sources. This appendix presents two circuits the author finds most useful.

Voltage Difference Constant Current Source

Figure C-1 illustrates a basic voltage difference constant current source.

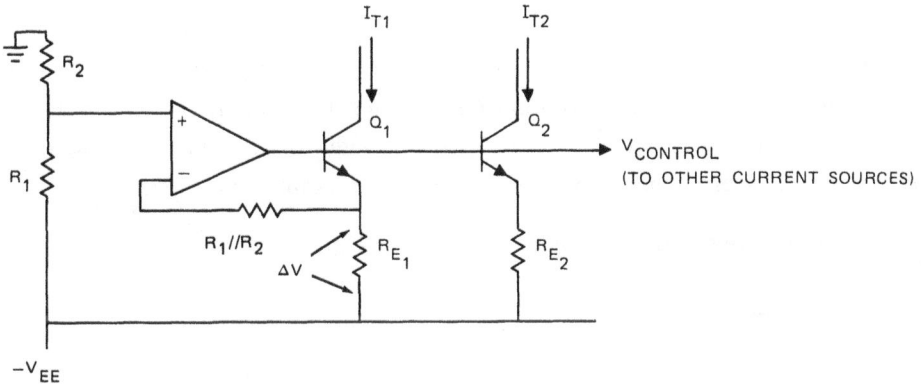

FIGURE C-1. Voltage Difference Constant Current Source.

The current I_{T1} is controlled (as shown in Figure C-1) directly and may be given as

$$I_{T1} \cong \frac{\Delta V}{R_{E1}} \tag{C-1}$$

where

$$\Delta V = \frac{R_1}{R_1 + R_2} \, V_{EE} \tag{C-2}$$

Current I_{T2} may be given as

$$I_{T2} = I_{T1}\left(\frac{R_{E1}}{R_{E2}}\right) - \frac{\Delta V_{BE}}{R_{E2}}$$ (C-3)

where

$$\Delta V_{BE} = V_{BE2} - V_{BE1}$$

If $R_{E1} \cong R_{E2}$, the two currents are quite close for well-matched transistors.

Due to feedback action, the current control voltage, ΔV, is insensitive to temperature variations of the base-emitter voltage. Thus I_{T1} and I_{T2} are quite insensitive to temperature variations (provided $R_{E2} \cong R_{E1}$). If the transistors and current determining resistors are contained in an integrated circuit, this bias technique will be quite temperature sensitive due to the change in absolute resistance values with temperature (see Equation C-1).

Temperature Insensitive Current Source

The absolute resistance value for integrated circuit resistors can vary greatly with temperature and from chip-to-chip (see Chapter 4); however, the resistance ratio over temperature and from chip-to-chip is quite close. The current source illustrated in Figure C-2 takes advantage of this temperature insensitive ratio.

The temperature insensitive reference current, I_{REF}, is given as

$$I_{ref} = \frac{\Delta V}{R_{REF}}$$ (C-4)

where

$$\Delta V = \frac{R_1 V_{CC}}{R_1 + R_2}$$ (C-5)

Neglecting any base currents (Q_2 provides base drive current to the constant current transistors Q_1, Q_3, and Q_4), the collector current of Q_1 is

$$I_{C1} = I_{REF}$$ (C-6)

136

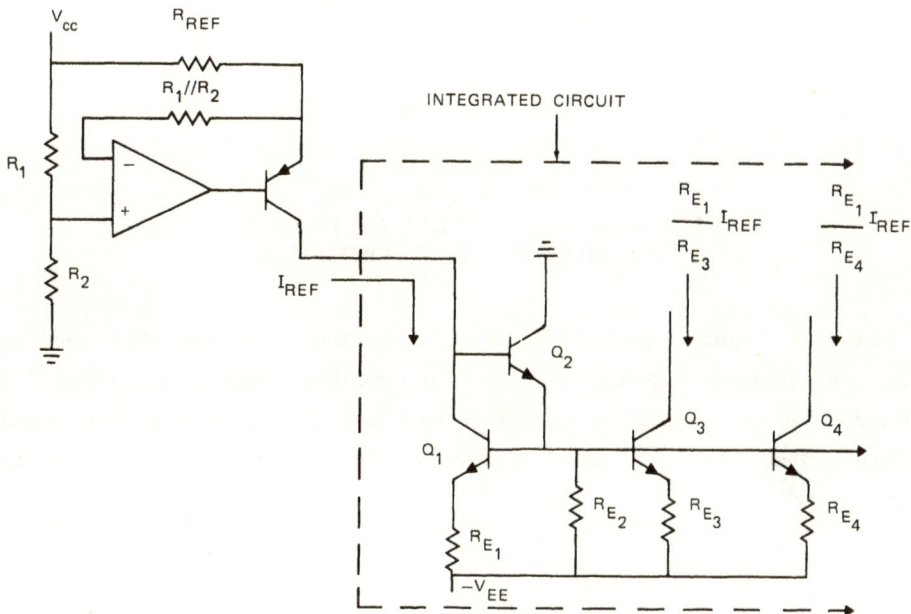

FIGURE C-2. Temperature Insensitive Current Source.

Assuming well-matched transistors

$$I_{C3} = \frac{R_{E1}}{R_{E2}} I_{REF} \qquad (C\text{-}7)$$

Thus, the constant current sources are accurately controlled by the ratio of R_{E1}/R_{En} [1], and this ratio is temperature and device-to-device insensitive.

References

1. Gray, P. R. and R. G. Meyer. *Analysis and Design of Analog Integrated Circuits.* 2nd ed. New York: John Wiley and Sons, 1984.

Appendix D

OUTPUT STAGES USEFUL FOR DETECTOR-
LOGARITHMIC VIDEO AMPLIFIERS

Chapters 2 and 4 present the parallel summation logarithmic video amplifier using the differential amplifier as the video logarithmic element, and Figure D-1 illustrates the basic logarithmic element configuration. This appendix presents several techniques that have proven useful in delivering the logarithmic video output to the outside world.

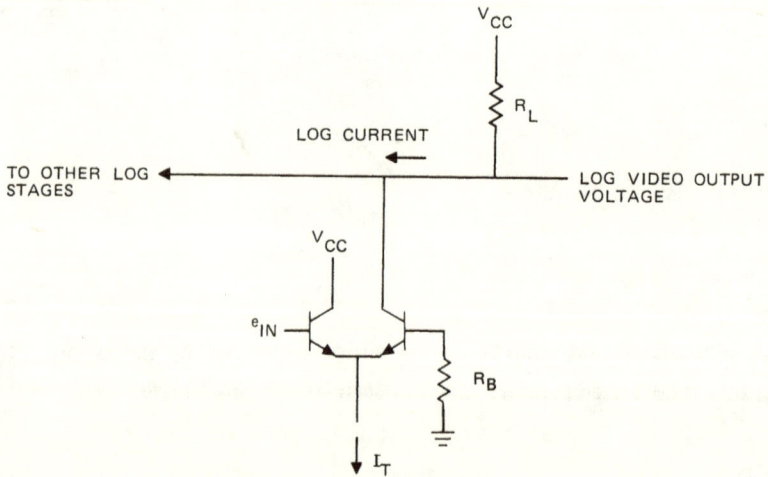

FIGURE D-1. Basic Logarithmic Element.

To remove R_L from the parallel collector capacitance of the logging transistors, a common base (or cascode) transistor is placed between R_L and the logarithmic elements, as shown in Figure D-2. If the load can be capacitively coupled, the DC offset, due to the bias of the logarithmic elements, is not a problem and a simple emitter follower (Q_2) can be used as shown.

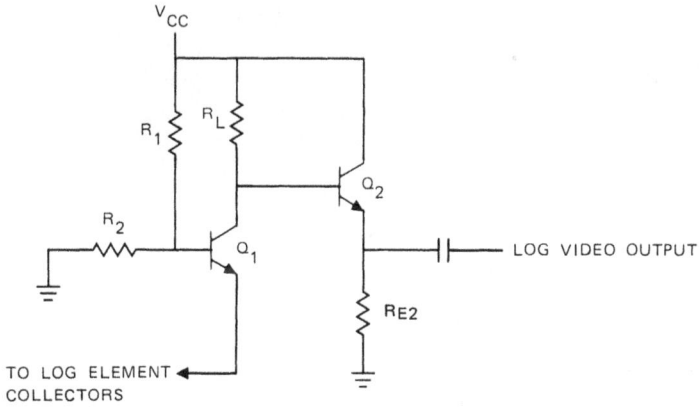

FIGURE D-2. Common Base (Q_1)/Emitter Follower (Q_2) Output Stage (see Figure 2-29a and -29b).

If the logarithmic video must be DC coupled to the load, a DC-level shifting stage is necessary (see Figure 4-11). Figure D-3 illustrates a simple DC-level shifting output stage that is easily realized with integrated circuit components. The DC collector voltage of Q_1 is due to the quiescent logarithmic element bias currents and is given as

$$V_{C1} = V_{CC} - \left(\sum \frac{I_T}{2} \right) R_L \tag{D-1}$$

where

$$\frac{I_T}{2} = \text{quiescent logarithmic element bias current}$$

The quiescent DC output voltage (V_{E4}) may be given as

$$V_{E4} = V_{C1} - V_{BE2} - I_{C3}R_{C3} - V_{BE4} \qquad \text{(D-2)}$$

Since

$$I_{C3} \cong I_{E3}$$

$$I_{C3} = \frac{V_B - V_{BE3} - V_{EE}}{R_{E3}} \qquad \text{(D-3)}$$

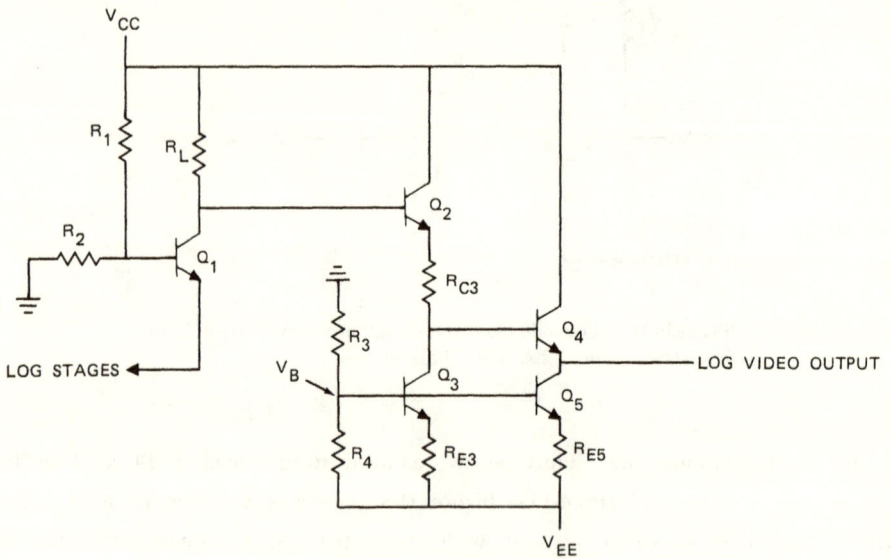

FIGURE D-3. DC Level Shifting Output Stage (see Figure 4-11).

The DC output voltage, V_{E4}, is now given as

$$V_{E4} = V_{C1} - V_{BE2} - \frac{(V_B - V_{BE3} - V_{EE})\, R_{C3}}{R_{E3}} - V_{BE4} \qquad \text{(D-4)}$$

or

$$V_{E4} = V_{C1} + \frac{R_{C3}}{R_{E3}} (V_{EE} - V_B) - V_{BE4} - V_{BE2} + \frac{R_{C3}}{R_{E3}} V_{BE3} \qquad \text{(D-5)}$$

If $R_{E5} = R_{E3}$, Q_2, Q_3, and Q_4 have the same emitter current; thus, $V_{BE4} = V_{BE2} = V_{BE3}$ and Equation D-5 may be given as

$$V_{E4} = V_{C1} + \frac{R_{C3}}{R_{E3}} (V_{EE} - V_B) - 2V_{BE} + \frac{R_{C3}}{R_{E3}} V_{BE} \qquad \text{(D-6)}$$

The only temperature-dependent term in Equation D-6 is V_{BE} (for well-designed logarithmic stages, I_T will be temperature-insensitive, Equation D-1, and I_{C1} thus V_{C1}, is temperature-insensitive). If

$$\frac{R_{C3}}{R_{E3}} = 2 \qquad \text{(D-7)}$$

then the V_{BE} terms cancel, and V_{E4} is temperature-insensitive. The necessary bias voltage, V_B, to give $V_{E4} = 0$, can now be found as

$$V_B = V_{EE} + \frac{V_{C1}}{2} \quad (V_{E4} = 0 \text{ volts}) \qquad \text{(D-8)}$$

Chapter 4, Figure 4-12, presents a DC feedback technique to minimize the DC offset voltage associated with the two linear video amplifiers. If this technique can be used, the circuit shown in Figure D-4 will ensure the quiescent logarithmic output voltage is 0 V. Transistor Q_3 provides the necessary logarithmic bias current $\left(\sum \frac{I_T}{2} \right)$, the emitter bias current for Q_1 (the common base, or cascode, transistor), and the emitter follower Q_2. Feedback action ensures that the quiescent logarithmic video output voltage is 0 V, independent of any changes in the logarthmic bias currents with temperature. When a signal is present, the signal threshold comparator opens the switch (thus removing the signal from the DC nulling loop). The DC output voltage for this technique is simply the offset voltage of the operational amplifier plus the op-amp offset bias current times R. Maximum output voltages less than 5 mV are simple to achieve.

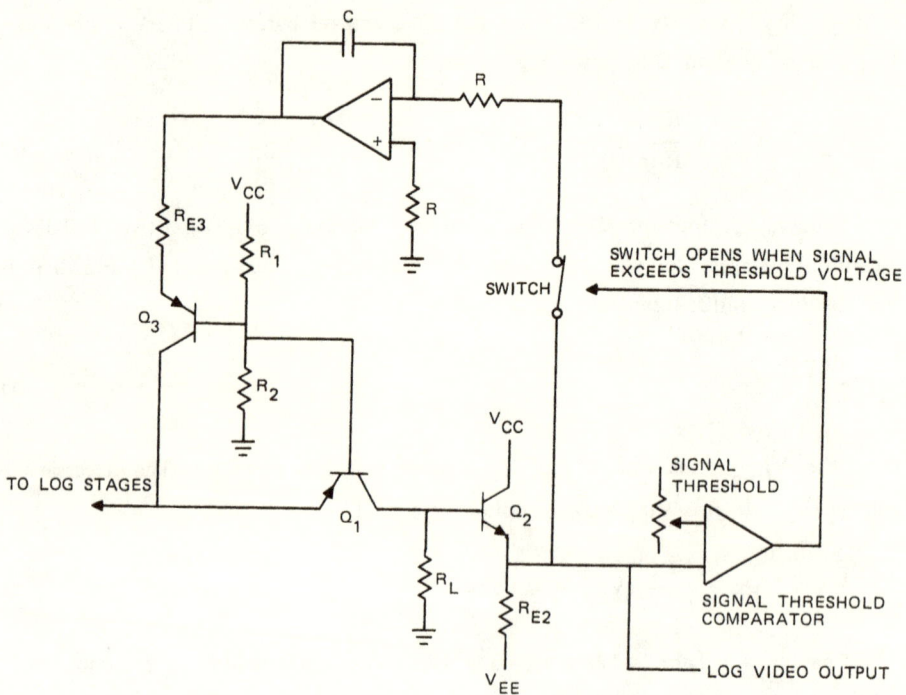

FIGURE D-4. Logarithmic Output D C Nulling Technique.

142

Appendix E

SELECTED LINEAR VIDEO AMPLIFIERS

The two basic types of linear video amplification, presented in Chapters 2 and 4, are (1) nonlimiting, wide bandwidth, ultralow DC-offset and drift and (2) limiting, with wide bandwidth and low DC offset. The author knows of no commercial integrated circuit operational amplifiers that fulfill the above needs. This appendix presents several circuit approaches that have proven useful and a summary of a design currently being designed into a master-chip integrated circuit.

Nonlimiting Amplifier

Figure E-1 illustrates the basic detector/linear video approach presented in Chapters 2 and 4. This amplifier is not driven into saturation, so pulse recovery time problems are minimized. The DC offset and drift needed (Chapter 4) are exceptionally low, however. At present, the author knows of no commercial amplifier that posesses the necessary bandwidth and DC drift characteristics suitable for use with detector-logarithmic video amplifiers. The amplifier illustrated in Figure E-2 (and discussed in detail in Chapter 4) has been used with success; however, the DC offset adjustment may have to be made after temperature drift data has been taken for the individual amplifier. It is possible that several iterations may be needed before the optimum DC offset nulling can be obtained. The supermatched transistors, Q_1 (A and B) may not lend themselves to integration with the logarithmic elements; however, transistors Q_2 through Q_6 and their associated resistors may easily be integrated. Several commercial logarithmic amplifier and operational amplifier manufacturers are getting quite close to practical monolithic designs, and it is just a matter of time before practical units become available.

143

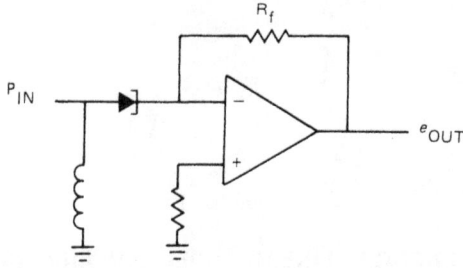

FIGURE E-1. Basic Detector/Linear
Video Amplifier.

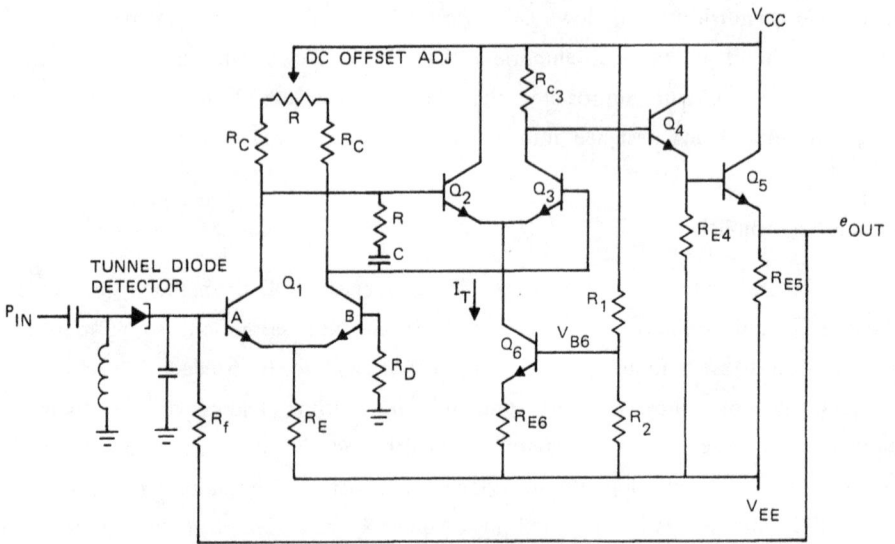

FIGURE E-2. D.C. Stable Video Amplifier Using Supermatched
Transistors (Q_1A and Q_2B) (see Chapter 4).

Linear-Limiting Amplifier

The other linear amplifier in the logarithmic video amplifier discussed in Chapters 2 and 4 must limit to prevent overdriving the emitter-base junction of the lowest level logarithmic element. If this amplifier is driven into saturation, severe recovery time occurs. The amplifier of Figure E-1 can be easily converted to a

144

noninverting-limiting amplifier, as discussed in Chapter 4 (see Figure 4-9). An alternate approach is illustrated in Figure E-3 (also see Figure 2-35). Transistors Q_1 (a supermatched pair) and Q_6 comprise a simple differential amplifier (see Appendix A). The bias current for Q_1 (I_T) is provided by Q_6, Q_7, and I_{REF} (see Appendix C, Figure C-2), and is given as

$$I_T = \frac{R_K}{R_{E1}} I_{REF} \tag{E-1}$$

I_T is temperature-insensitive, thus the collector voltage of Q_B,

$$V_{CB} = V_{CC} - \frac{I_T R_C}{2} \tag{E-2}$$

is temperature-independent. Transistors Q_2 through Q_5 and their associated resistors comprise a temperature stable level shifting amplifier (see Appendix D, Figure D-3) and is designed to ensure a 0–VDC output with a 0-V input.

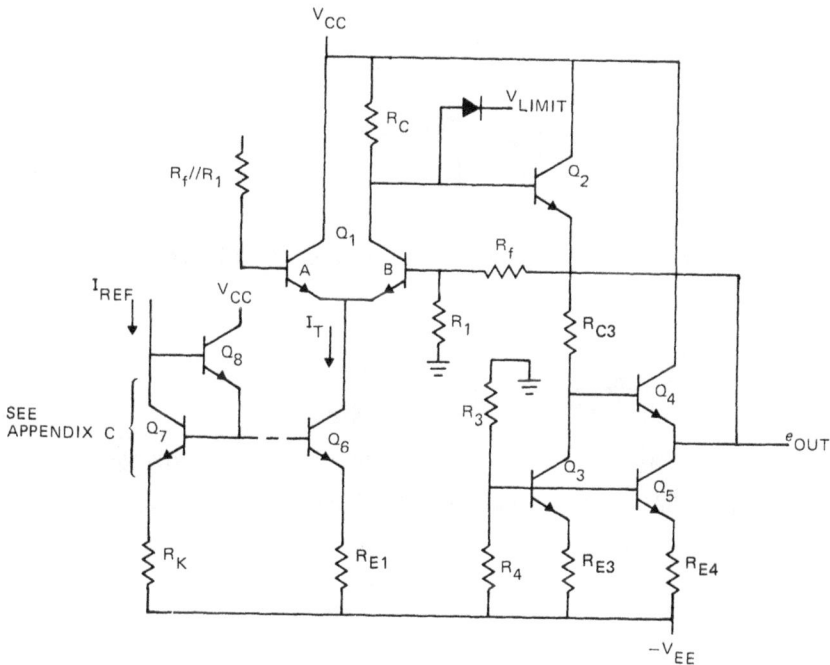

FIGURE E-3. Lin-Limit Feedback Video Amplifier.

145

The amplifier of Figure E-3 can be represented as shown in Figure E-4. The voltage gain, by classical feedback theory, is

$$A_V = \frac{A_o}{1 + BA_o} \qquad \text{(E-3)}$$

where

$$B = \frac{1}{1 + R_f/R_1}$$

A_o = low frequency open loop voltage gain

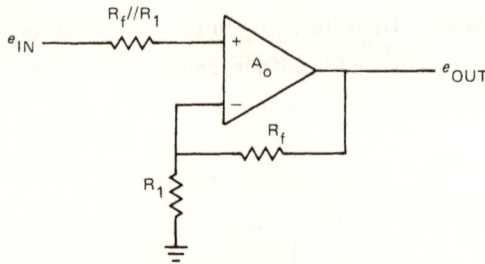

FIGURE E-4. Noninverting Feedback Video Amplifier.

Generally, BA_o (the loop gain) is considered large with respect to 1, and the closed loop voltage gain reduces to

$$A_V = \frac{R_f + R_1}{R_1} \quad \text{(loop gain} \gg 1) \qquad \text{(E-4)}$$

Unfortunately, A_o for the circuit of Figure E-1 is relatively small, and Equation E-4 is not valid. The open loop gain for Figure E-1 (assuming the voltage gain of Q_2 and Q_4 is unity) is given as

$$A_o \cong \frac{R_C}{r_e'A + r_e'B} \qquad \text{(E-5)}$$

146

where the dynamic emitter resistance, r_e', is given as

$$r_e' = \frac{V_T}{I_E} \qquad\qquad (E\text{-}6)$$

where

$$V_T = \frac{KT}{q} = \frac{86.25 \times 10^{-6}\ T}{I_E} \qquad\qquad (E\text{-}7)$$

and

$$I_E = \frac{I_T}{2} \qquad\qquad (E\text{-}8)$$

Substituting Equations E-8 and E-7 into E-5

$$A_o \cong \frac{I_T\ R_C}{345 \times 10^{-6}\ T} \qquad\qquad (E\text{-}9)$$

It will be shown that the gain, A_o, is maximized by decreasing the collector voltage of Q_{1B}, V_{CB}.

$$V_{CB} = V_{CC} - I_{CB}\ R_C \qquad\qquad (E\text{-}10)$$

$$I_{CB} = \frac{I_T}{2} \qquad\qquad (E\text{-}11)$$

thus

$$V_{CB} = V_{CC} - \frac{I_T}{2}\ R_C \qquad\qquad (E\text{-}12)$$

However, I_T, from Equation E-9, is given as

$$I_T = \frac{345 \times 10^{-6}\ T}{R_C}\ A_o \qquad\qquad (E\text{-}13)$$

Substituting Equation E-13 into E-12 and solving for A_o

$$A_o = \frac{V_{CC} - V_{CB}}{172 \times 10^{-6}\ T} \qquad\qquad (E\text{-}14)$$

147

Thus, decreasing V_{CB} increases the voltage gain.

Table E-1 gives the open loop gain, A_o, as a function of temperature for $V_{CC} = +12$ V, and $V_{CB} = +2$ V.

TABLE E-1. A_o as a Function of T
($V_{CC} = +12$ V, $V_{CB} = +2$ V).

T	A_o (dB)	T (°C)
223	261 (48.3)	–50
300	193 (45.7)	+27
373	156 (43.8)	+100

The effect of this temperature dependent gain on a practical 30-dB (31.63) amplifier will now be found.

The closed loop voltage gain for the circuit of Figure E-2 is given as

$$A_V = \frac{A_o}{1 + BA_o} \tag{E-15}$$

Solving Equation E-15 for B

$$B = \frac{\dfrac{A_o}{A_V} - 1}{A_o} \tag{E-16}$$

The value of B needed for a 31.63 gain (30 dB) at 27°C (300°K) is

$$B\big|_{300°K} = \frac{\dfrac{193}{31.63} - 1}{193} = 0.0264 \tag{E-17}$$

The ratio of R_f/R_1 as a function of B is given as

$$\frac{R_f}{R_1} = \frac{1}{B} - 1 \tag{E-18}$$

or

$$\frac{R_f}{R_1} = 36.83 \hspace{4cm} \text{(E-19)}$$

Table E-2 shows the closed loop gain, Equation E-3, as a function of temperature. It can be seen that the closed loop voltage gain changes by less than ±0.5 dB with temperature. This temperature dependent gain error will cause a logarithmic conformity error for low-level signals (see Chapter 4 and Appendix B). Individual applications determine if this error can be tolerated; however, this amplifier configuration is quite simple.

TABLE E-2. A_V as a Function of T.

T	A_O (Table E-1)	B (Equation E-18)	A_V (dB) (Equation E-3)	T (°C)
223	261	0.0264	33 (30.4)	–50
300	193	0.0264	31.66 (30)	+27
373	156	0.0264	30.5 (29.7)	+100

Figure E-5 illustrates a 30-dB lin-limiting feedback video amplifier. A voltage difference constant current source (see Appendix C) is used to provide I_T (5.55 mA). With the zero adjust potentiometer set for a 2.00 V collector voltage (I_T = 5.55 mA), the DC output voltage was +12 mV. When the output was adjusted to 0 volts, it drifted less than ±2 mV with temperature. The voltage gain was as predicted in Table E-2. If this stage is used to drive the two low-level logarithmic stages, a simple 5-V zener diode can be used (see Figure 2-36) as the limiting voltage, V_L.

Figure E-6 illustrates the pulse response for this amplifier.

The amplifier configuration of Figure E-3 is ideal for logarithmic video application due to its inherent limiting for large input signals and its ease to varying the limiting level (V_{limit}). The primary disadvantage is the limited open loop voltage gain (see Equation E-14). The open loop voltage gain can be substantially increased if R_C in Figure E-3 is replaced by an active load.

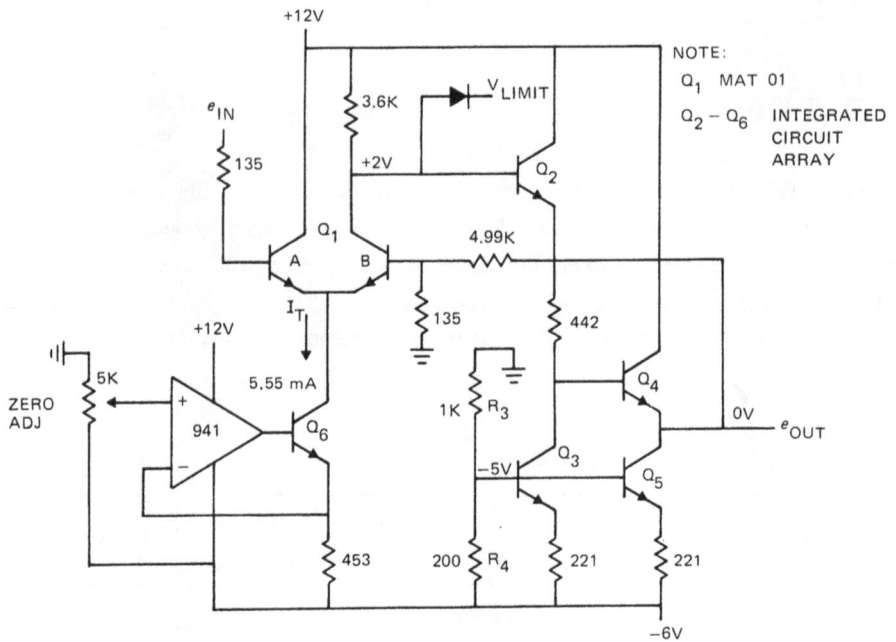

FIGURE E-5. 30-dB Lin-Limit Feedback Video Amplifier.

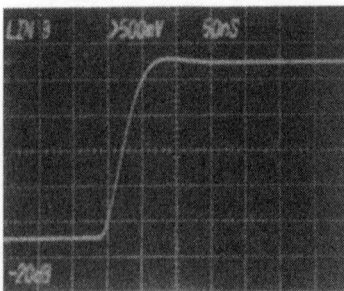

(a) Input = -20 dB (0.1 V).

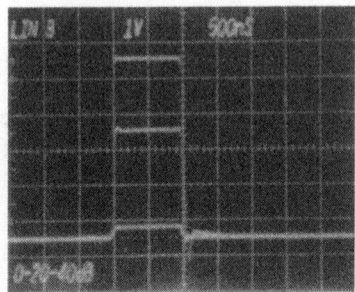

(b) Inputs = 0, -20, -40 dB.

FIGURE E-6. Pulse Response for the 30-dB Lin-Limit Video
Amplifier (see Figure E-5).

150

Linear-Limiting Amplifier – A Basic
Integrated Circuit Approach

Figure E-7 illustrates the basic integrated circuit lin-limit video amplifier. Transistors Q_1, Q_2, and Q_3 comprise the basic differential amplifier, with Q_4 comprising an active load [1]. Open loop gains in excess of 60 dB are possible with this configuration. To minimize bias current shifts with variations in V_{CC} and V_{EE}, a voltage independent current source is used to generate the reference current, I_{REF} [2]. The DC level shifting is described in Appendix D (Figure D-3). The limiting voltage is a temperature compensated zener diode.

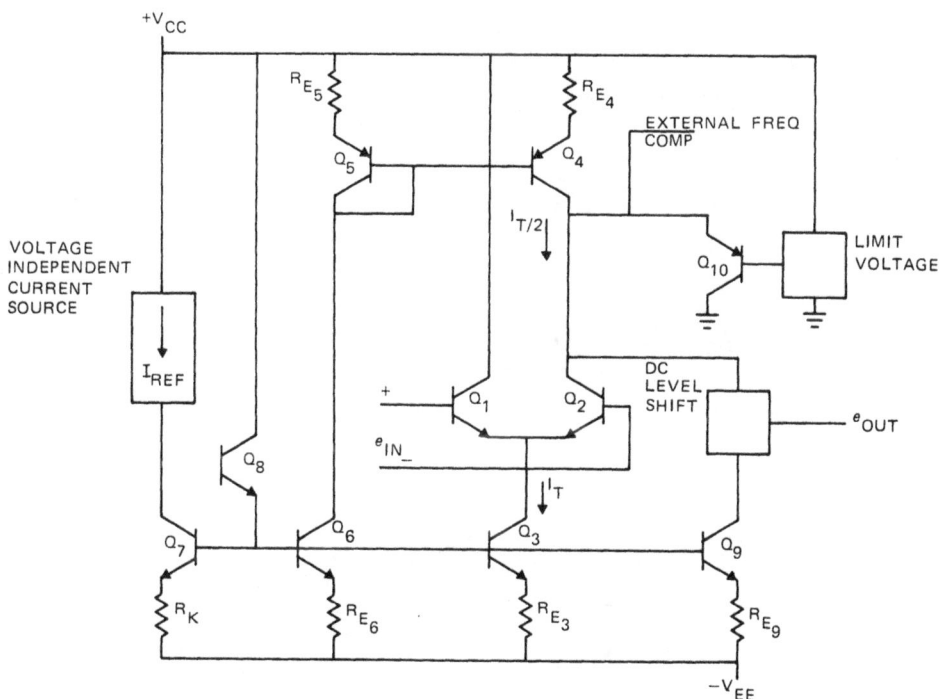

FIGURE E-7. Basic Integrated Lin-Limit Video Amplifier.

151

This configuration is suitable for both the detector video amplifier and the 30-dB noninverting amplifier; however, the temperature drift exceeds the 1.3 $\mu V/°C$ (see Equation E-4 and E-5) necessary for the detector amplifier. Supermatched transistor techniques [3] can be used to reduce the V_{BE} mismatch, but this will result in a slower amplifier. I fear the answer lies with a custom integrated circuit. If the reader has a technique suitable for the master chip approach, it is certainly worth publishing, so we can all benefit.

References

1. Gray, P. and R. Meyer, *Analysis and Design of Analog Integrated Circuits.* New York: John Wiley and Sons, 1977, pp. 210–217.

2. Ibid. pp. 239–248.

3. Interdesign, Inc. *Super-Matched Differential Pair*, by D. Bray and E. MacPherson. (Monochip Application Note APN-20.)

Appendix F

BASIC DETECTORS

Logarithmic amplifiers, whether logarithmic IF or detector-logarithmic videos, rely on the detector to convert the IF (or RF) signal into a voltage suitable for signal processing. This appendix reviews the basic operation of the Schottky diode and tunnel diode detectors. It will be shown that if one is interested in "true" DC-coupled detector-logarithmic amplification, the tunnel diode may well be the only practical choice.

Schottky Diode Detector

The general curve relating output voltage to input power (in dBm) is shown in Figure F-1. The pulse output increases 2 dB for every 1-dB increase in input power (square law region) for inputs up to approximately -20 dBm. Above -20 dBm the detector starts acting as a peak detector such that for inputs larger than 0 dBm the detector is in the linear range with the pulse output increasing 1 dB for each dBm increase in input power.

One useful Schottky diode parameter is the diode K [1], where

$$K \equiv \frac{mV \ output}{mW \ input} \tag{F-1}$$

or, in terms of dBm

$$K \equiv \frac{e_d(in \ mV) \ mV}{10 \times exp\left(\frac{P_{dBm}}{10}\right) mW} \tag{F-2}$$

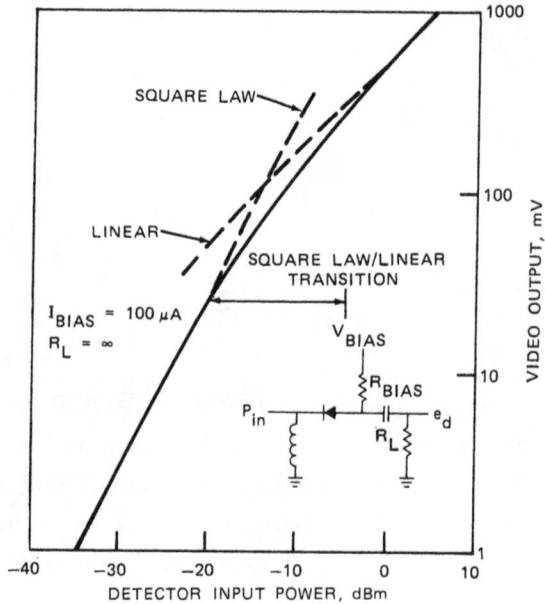

FIGURE F-1. Schottky Detector Characteristic.

and K may be thought of as a static gain (e_0 in mV for a given input in mW). The K is a function of the detector's bias current, load resistance, input power, temperature, and input frequency. Table F-1 lists the measured video outputs for various input powers, bias currents, and load resistors. (The K, Equation F-2, for each condition is given in parentheses.) Figure F-2 is a plot of e_0 versus P_{in}, and Figure F-3 is a plot of K versus P_{in}. It can be seen from Table F-1 and Figures F-2 and F-3 that bias has a fairly strong influence on K for the unloaded detector and much less an effect for $R_L = 200\ \Omega$. (Loading a Schottky diode to improve the video rise time is a common practice (see [1], [2]).

It is also seen from Figure F-3 that for inputs below −22 to −23 dBm the values of K are constant (square law), and that K decreases linearly (log scale) with input power in dB (linear law) above 0 dBm or so.

To be useful in high-density environments, the detector should be direct-coupled to its load resistor and to any following amplifier. This is a most difficult problem; one solution is illustrated in Figure F-4.

154

TABLE F-1. Detector Output, e_d(in mV) and Detector (K) as a Function of Bias, R_L, and P_{in}.

$I_{BIAS} = 50 \mu A$ ($V_F = 0.3312$)

R_L P_{in} (dBm)	∞	1000	500	200
-25	8.8 (2783)	5.8 (1834)	4.2 (1328)	2.5 (790)
-20	24.5 (2450)	16 (1600)	12 (1200)	7 (700)
-15	64 (2023)	42 (1328)	33 (1043)	21 (664)
-10	140 (1400)	98 (980)	80 (800)	52 (520)
-5	295 (933)	210 (664)	172 (544)	125 (395)
0	580 (580)	420 (420)	350 (350)	260 (260)
+5	1080 (342)	800 (253)	680 (215)	510 (161)

$I_{BIAS} = 75 \mu A$ ($V_F = 0.3425$)

R_L P_{in} (dBm)	∞	1000	500	200
-25	8 (2529)	6 (1897)	4.7 (1486)	3 (948)
-20	22.5 (2250)	16 (1600)	12.5 (1250)	8.2 (820)
-15	59.5 (1881)	42 (1328)	34 (1075)	22.5 (711)
-10	136 (1360)	96 (960)	79 (790)	54 (540)
-5	287 (908)	210 (664)	175 (553)	125 (395)
0	560 (560)	410 (410)	350 (350)	260 (260)
+5	1060 (335)	780 (247)	680 (215)	510 (161)

$I_{BIAS} = 100 \mu A$ ($V_F = 0.3502$)

R_L P_{in} (dBm)	∞	1000	500	200
-25	7 (2213)	5.2 (1644)	4.5 (1423)	3 (948)
-20	20 (2000)	15.2 (1520)	12.5 (1250)	8.2 (820)
-15	56 (1770)	41 (1296)	34 (1075)	23 (727)
-10	128 (1280)	95 (950)	79 (790)	56 (560)
-5	275 (870)	205 (648)	175 (553)	125 (395)
0	550 (550)	410 (410)	350 (350)	260 (260)
+5	1040 (329)	780 (247)	670 (212)	500 (158)

$I_{BIAS} = 150 \mu A$ ($V_F = 0.3618$)

R_L P_{in} (dBm)	∞	1000	500	200
-25	6 (1897)	5 (1581)	4.2 (1328)	3 (948)
-20	17.5 (1750)	14.5 (1450)	12.2 (1220)	8.5 (850)
-15	50 (1581)	40 (1264)	34 (1075)	24 (758)
-10	120 (1200)	94 (940)	80 (800)	57 (570)
-5	265 (838)	205 (648)	175 (553)	128 (405)
0	530 (530)	410 (410)	350 (350)	260 (260)
+5	1020 (323)	780 (247)	670 (212)	500 (158)

155

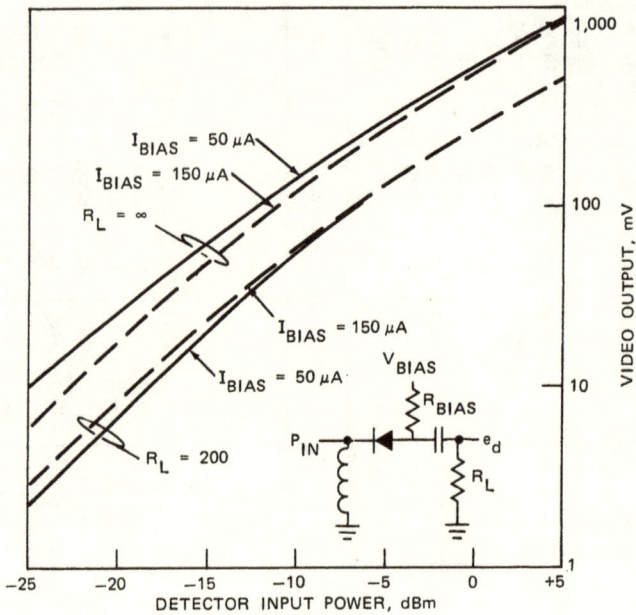

FIGURE F-2. Video Output Versus Detector Input Power.

The concept here depends on the virtual ground at the inverting input of the operational amplifier ($V_- = V_+$).

Thus

$$ID_2 = \frac{V_{BIAS} - VD_2}{R_2 + R_{BIAS}} \tag{F-3}$$

$$V_+ = ID_2R_2 + VD_2 \tag{F-4}$$

Since

$$V_- = V_+ \tag{F-5}$$

$$ID_1 = \frac{ID_2R_2 + VD_2 - VD_1}{R_1} \tag{F-6}$$

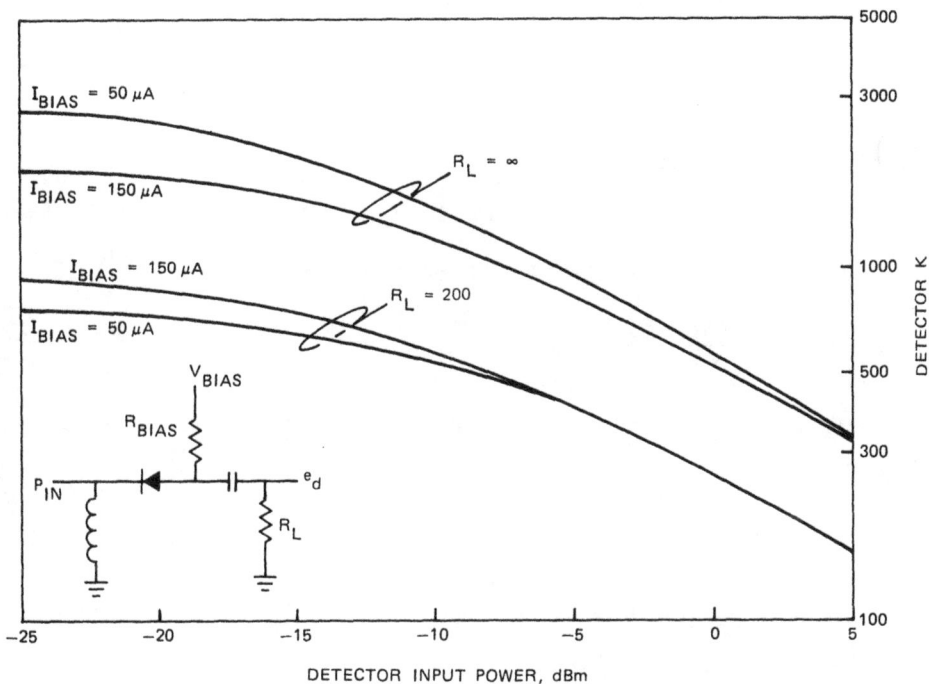

FIGURE F-3. K Versus Input Power.

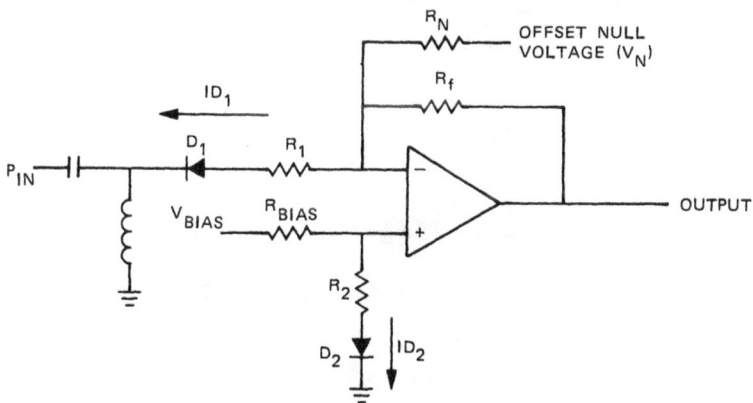

FIGURE F-4. Direct-Coupled Schottky Diode Detector/Video Amplifier.

157

or

$$ID_1 = ID_2 \left(\frac{R_2}{R_1}\right) + \frac{\Delta VD}{R_1} \qquad \text{(F-7)}$$

where

$$\Delta VD = \text{diode DC voltage match } (VD_2 - VD_1) \qquad \text{(F-8)}$$

If $R_1 = R_2$, the detector bias current, ID_1, becomes

$$ID_1 = ID_2 + \frac{\Delta VD}{R_1} \qquad \text{(F-9)}$$

Since ID_2 can be held quite constant, the change in the detector bias current, ΔID_1, is

$$\Delta ID_1 = \frac{\Delta VD}{R_1} \qquad \text{(F-10)}$$

Table F-2 gives the change in detector bias current for various diode matches for $R_1 = 200$.

TABLE F-2. Effect of DC Diode
Match on Bias Current.
($R_1 = R_2 = 200 \ \Omega$)

ΔVD, mV	ΔID_1, μA
±1	±5
±5	±25
±10	±50
±15	±75

Referring to Figure F-2, it is seen that an $R_L = 200$ and a bias current of 50 μA results in approximately a 1-dB error with respect to a bias current of 150 μA. Thus, if two matched detectors/amplifiers were needed and one had a 150-μA bias and the other a 50-μA bias, the output pulse amplitude difference would be 1 dB referred to the input. This is generally excessive; thus the DC voltage match, ΔVD, should be within ±5 mV or better.

158

The DC temperature shift for Schottky diodes biased at 100 μA is

$$\frac{\Delta V_F}{\Delta T} \cong -1.5 \text{ mV/}^\circ\text{C} \tag{F-11}$$

where

$\Delta V_F \equiv$ change in DC voltage drop with temperature

Thus, to ensure equal bias currents for a matched pair of detector/video amplifiers, D_1 and D_2 not only must have $\Delta VD \leqslant 5$ mV, they must also have the same temperature coefficient and be thermally close.

There is another inherent difficulty (besides the DC-matched diodes) with the concept of Figure F-4. There is a DC offset voltage at the output of the op-amp (the output must supply the bias current for D_1). The op-amp DC output may be given as [3]

$$V_o = ID_2R_2 \left(\frac{R_1 + R_f}{R_1}\right) + VD_2 + \frac{R_f}{R_1} \Delta VD \tag{F-12}$$

and this voltage can be set to zero by way of the offset null voltage. However, this "zero" state is only valid for one temperature, as VD_2 in Equation F-12 is temperature-dependent (see Equation F-11). Thus, assuming DC nulling at one temperature, the change in output voltage is

$$\Delta V_o/\Delta T = \Delta V_F D_2/\Delta T \tag{F-13}$$

or

$$\Delta V_o \cong -1.5 \times 10^{-3} (\Delta T) \tag{F-14}$$

where

$$\Delta T = T_2 - T_1 \tag{F-15}$$

Thus, for

$T_1 = 25^\circ\text{C},$

$$\Delta V_o|_{T_2 = -40^\circ\text{C}} = +97.5 \text{ mV} \tag{F-15a}$$

159

$$\Delta V_o \big|_{T_2 = +80°C} = -82.5 \text{ mV} \qquad\qquad\text{(F-15b)}$$

and this temperature shift will be excessive if true DC-coupled gain must follow this amplifier, as is the case when this circuit is used as the detector for true DC-coupled logarithmic amplifiers. Thus, no matter how close the match (and temperature tracking) of VD_1 and VD_2, a DC null voltage is necessary to keep V_o (DC) = 0 as VD_2 changes its −1.5 mV/°C with temperature. If pseudo−DC coupling (see Chapter 4, Figure 4-12) can be used, the Schottky detector configuration, Figure F-4, can be used. However, if matched detector-logarithmic amps are needed, the Schottky detector poses additional problems, as will be shown.

The video (or pulse) output for the amplifier of Figure F-4 will now be found. From Equation F-2, the detector output is

$$e_d(\text{mV}) = K \ 10^{\frac{P_{dBm}}{10}} \qquad\qquad\text{(F-16)}$$

where it has been shown that K is a function of bias current, load resistance, and input power. Resistor R_1 in Figure F-4 is the load resistance for the detector, D_1. The detector ouput effectively drives the inverting input of the operational amplifier by way of R_1.

Thus

$$e_o = - \frac{R_f}{R_1} e_d \qquad\qquad\text{(F-17)}$$

or, using Equation F-16

$$e_o(\text{mV}) = - \frac{R_f}{R_1} K \ 10^{\frac{P|_{dBm}}{10}} \qquad\qquad\text{(F-18)}$$

Figure F-5 summarizes the theory presented thus far. To verify this theory, the configuration of Figure F-6 was built and tested. The results are listed in Table F-3, with the video output voltage versus input power illustrated in Figure F-7.

$$e_o \text{ (mV)} = -\frac{R_F}{R_1} K\, 10^{\frac{P_{dBm}}{10}}$$

$$ID_2 R_2\left(\frac{R_1 + R_f}{R_1}\right) + VD_2 + \frac{R_f}{R_1} \Delta VD$$

$$\left.\frac{\Delta V_O}{\Delta T}\right|_{IDEAL} = -1.5 \text{ mV/}^\circ C$$

$$ID_1 = \underbrace{\frac{V_{BIAS} - VD_2}{R_{BIAS} + R_2}}_{ID_2} + \frac{\Delta VD}{R_1}$$

$$\Delta VD = VD_2 - VD_1$$

NOTE: THIS SUMMARY NEGLECTS V_N ($R_N = \infty$).
SEE REFERENCE 3 FOR V_O AS A FUNCTION OF V_N AND R_N.

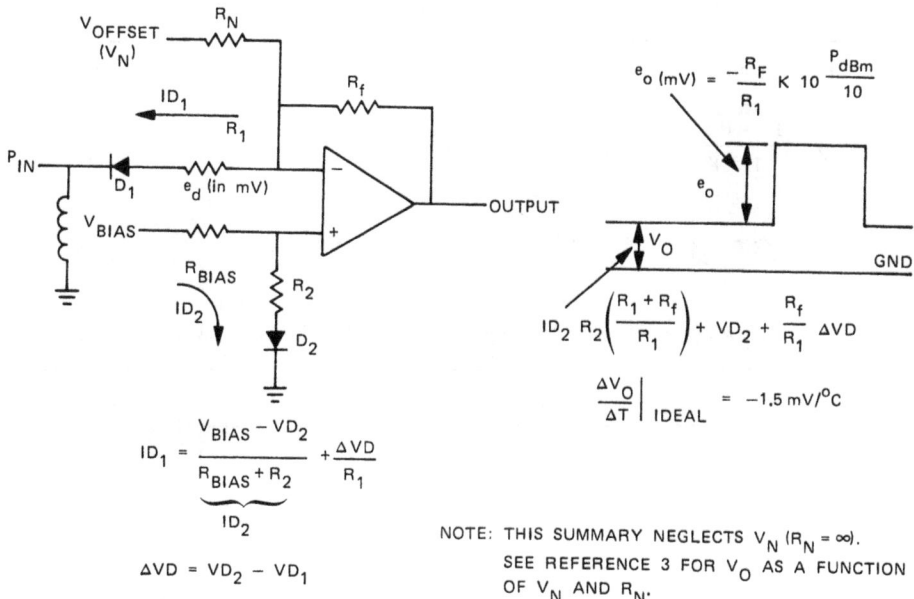

FIGURE F-5. Schottky Detector/Video Amplifier Summary.

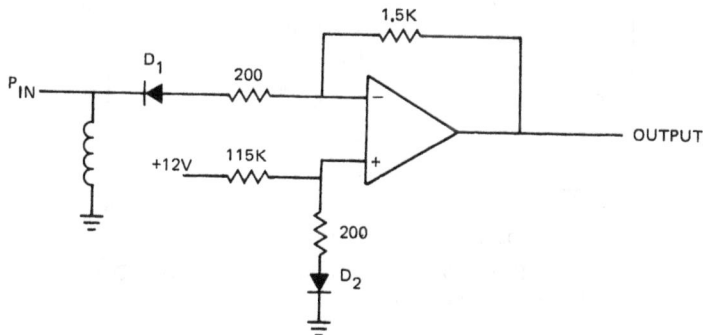

FIGURE F-6. Test Circuit.

TABLE F-3. Measured and Calculated Results for Figure F-6.

VD$_1$ = 0.346	VD$_2$ = 0.345	ΔV_F = -1 mV

$$ID_2 = \frac{12 - 0.345}{115 \text{ K}} = 101.3 \ \mu A \text{ (measured = 98.7 } \mu A)$$

$$ID_1 = ID_2 - \frac{1 \text{ mV}}{200} = 96.35 \ \mu A \text{ (measured = 92.7 } \mu A)$$

$$V_o = 101.3 \ \mu A \left(\frac{1.7 \text{ K}}{200}\right) + 0.345 - \frac{1.5 \text{ K}}{200} (1 \times 10^{-3}) = 509 \text{ mV (measured = 510 mV)}$$

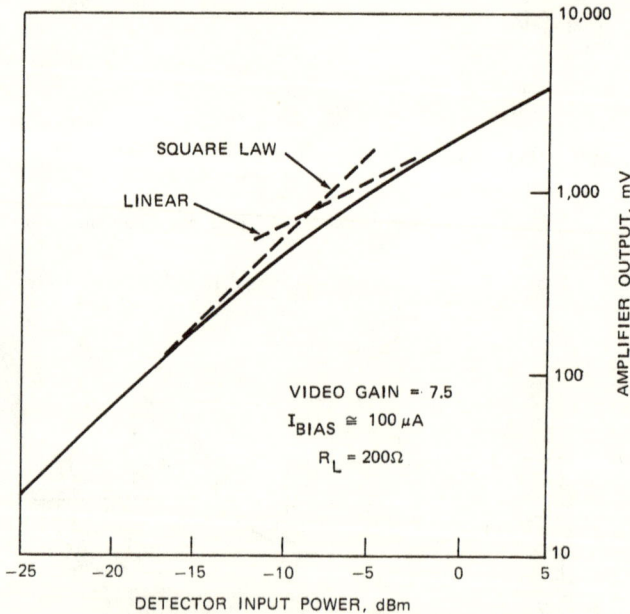

FIGURE F-7. Video Output Versus Input Power Test
Test Circuit (Figure F-6).

Many applications require two or more matched channels. This means that detector Ks must be matched, as will now be shown. Solving Equation F-18 for $P_{in} |_{dBm}$

$$P_{in}|_{dBm} = 10 \ \log \frac{R_1}{R_f} \ \frac{e_o(mV)}{K} \qquad (F\text{--}19)$$

Assume that two detectors/amplifiers have matched R_1 and R_f, and the input of one detector is varied until the pulse outputs are equal. The difference in input power in dBm (ΔP_{dB}) necessary to give equal video outputs is given as

$$\Delta P_{dB} = 10 \log \frac{K_A}{K_B} \qquad \text{(F-20)}$$

where

K_A = detector As K

K_B = detector Bs K

Solving Equation F-20 for the K match necessary for a given ΔP_{dB} match

$$\frac{K_A}{K_B} = 10^{\frac{\Delta P_{dB}}{10}} \qquad \text{(F-21)}$$

Table F-4 lists the K match necessary for given detector/amplifier matching, and, as seen, to obtain a match of, say, 0.5 dB, the K match must be within 12.2% over the input power range of interest, which places fairly severe DC-matching requirements on the DC reference diodes (see D_2 in Figure F-5). It must also be remembered that the DC output voltage will drift with temperature (no matter how good the diode match), and, to compensate for this drift, some form of continuous DC nulling (in the absence of input signals) will be required (see Figure 4-12).

TABLE F-4. K Match.

Input power match, ΔdBm	K_A/K_B	% K match
0.1	1.023	2.3
0.25	1.059	5.9
0.5	1.122	12.2
0.75	1.189	18.9
1.0	1.259	25.9

Figure F–8 summarizes the basics of Schottky diode detectors.

It is obvious that designing true DC–coupled stages with Schottky detectors is not at all a straightforward matter. There is a simpler solution to the problem,

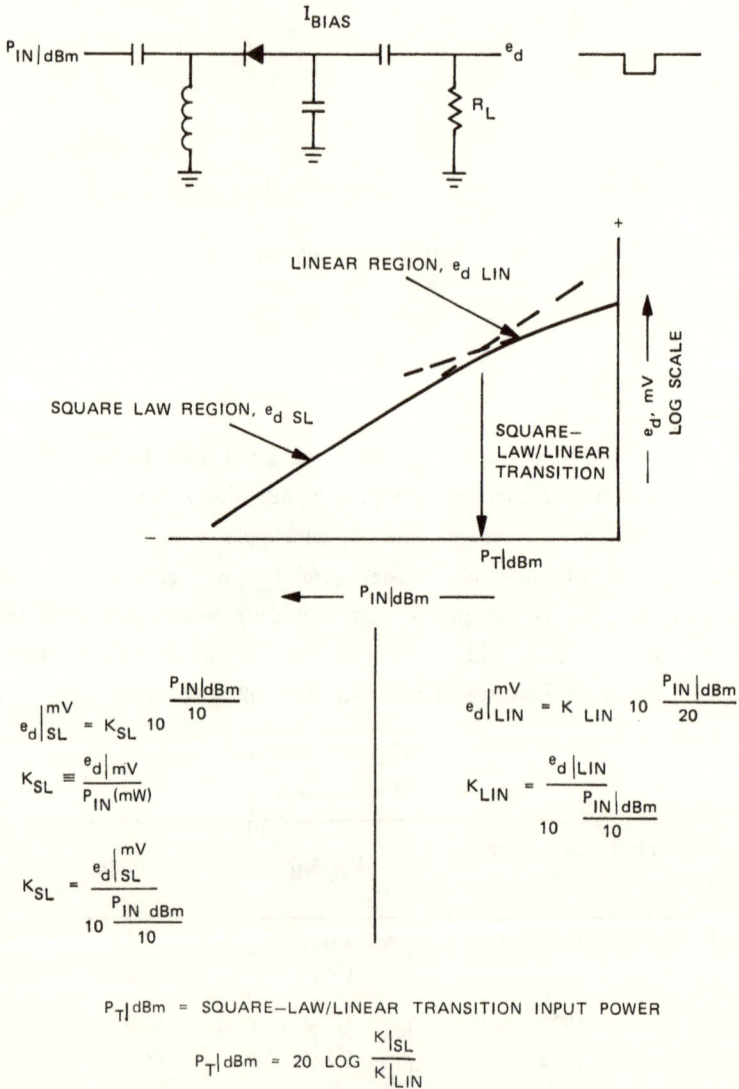

$$e_d\Big|_{SL}^{mV} = K_{SL}\ 10^{\frac{P_{IN}|dBm}{10}}$$

$$K_{SL} \equiv \frac{e_d|mV}{P_{IN}(mW)}$$

$$K_{SL} = \frac{e_d\Big|_{SL}^{mV}}{10^{\frac{P_{IN}\ dBm}{10}}}$$

$$e_d\Big|_{LIN}^{mV} = K_{LIN}\ 10^{\frac{P_{IN}|dBm}{20}}$$

$$K_{LIN} = \frac{e_d|LIN}{10^{\frac{P_{IN}|dBm}{10}}}$$

$P_{T}|dBm$ = SQUARE–LAW/LINEAR TRANSITION INPUT POWER

$$P_{T}|dBm = 20\ LOG\ \frac{K|_{SL}}{K|_{LIN}}$$

FIGURE F-8. Schottky Diode Detector Summary.

164

however, if the limited dynamic range and high temperature operation of tunnel diodes can be tolerated.

Tunnel Diode Detectors

Figure F-9 illustrates a simple and most useful tunnel diode detector/amplifier configuration [4]. The tunnel diode requires no DC bias and thus can be direct-coupled to the amplifier's inverting input.

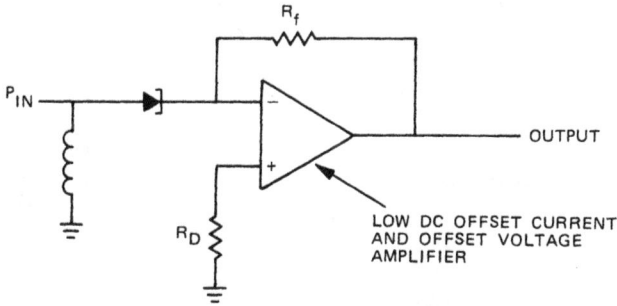

FIGURE F-9. Tunnel Diode Detector/Amplifier.

The DC output will be zero, in the absence of a signal, provided R_D is equal to the video resistance of the tunnel diode (R_V) (around 150 Ω). The pulse output may be given as

$$e_{out} = \beta R_f \left(10^{\dfrac{P_{in}|_{dBm}}{10}} \right) \times 10^{-3} \qquad \text{(F-22)}$$

where

$\beta \equiv$ diode short circuit current sensitivity

Figure F-10 illustrates the amplifier output versus detector input power. The diode–amplifier has a square law characteristic for inputs below –15 dBm or so. Above this level, the characteristic is nonlinear and never achieves linear detection.

165

The output limits for inputs greater than +5 dBm (the Schottky detector has a linear characteristic for inputs in excess of +15 dBm).

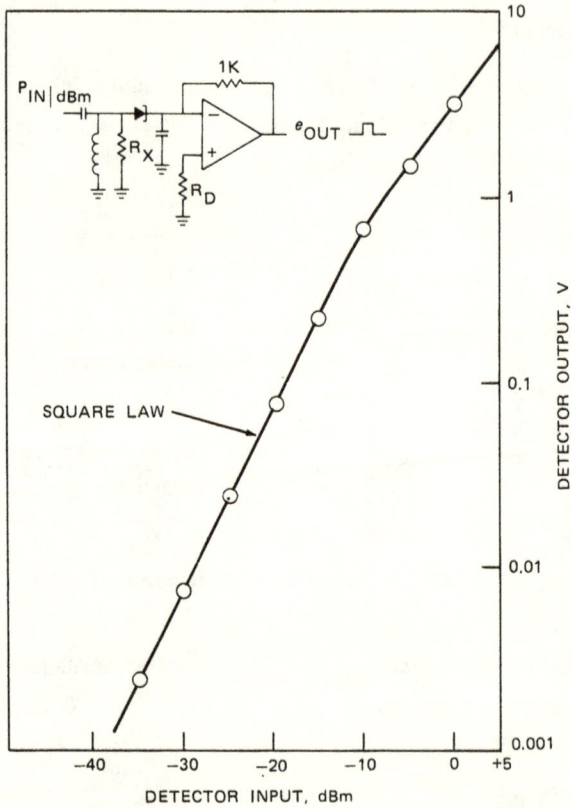

FIGURE F-10. Output Voltage Versus Detector Input Power.

Figure F-11 is a normalized plot (using the output at 0 dBm as the reference). It has been found experimentally that this normalized plot is a most reasonable approximation for any tunnel diode detector. The output for any R_f may be given as

$$e_{out}\Big|_{P_{in}}^{actual} = Z\ R_f\ (in\ k\Omega)\ e_{out}\ (obtained\ from\ Figure\ F\text{-}10) \qquad (F\text{-}23)$$

where

Z = measured output at 0 dBm and an R_f of 1 kΩ

166

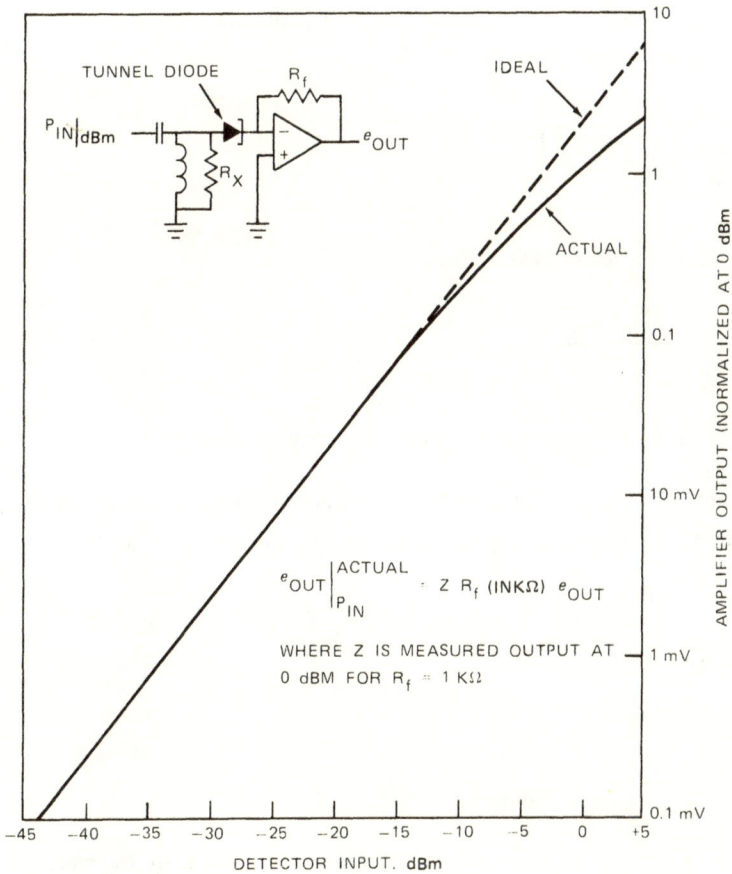

FIGURE F-11. Normalized Output Voltage Versus Detector
Input Power. (Values normalized at 0 dBm input.)

The detector configuration of Figure F-9 gives an inherent input VSWR of less than 2:1 for inputs from T_{ss} to +5 dBm. The input VSWR may be decreased (at the expense of T_{ss}) by R_x (see Figure F-10). R_x is chosen to give a 50-Ω input for a detector input around –15 dBm. The loss in T_{ss} is between 3 and 4 dBm; however, input VSWR less than 1.5:1 (over the full input dynamic range) are easily obtained.

167

Matched tunnel diode detectors are easily obtained by matching their βs. The detector β is obtained from Equation F-22 as

$$\beta = \frac{e_o}{R_f}\left(10^{\left(\frac{-P_{in|dBm}}{10}\right)}\right) \times 10^3 \tag{F-24}$$

If a value of $R_f = 1k\Omega$ is used

$$\beta = e_o \, 10^{\frac{-P_{in|dBm}}{10}}, \quad (R_f = 1k\Omega) \tag{F-25}$$

and since β is independent of R_f, this value may be used in your circuit.

To obtain matched detectors, it is necessary to match only the detector's βs. The β match necessary for a given input match may be given as [5]

$$\frac{\beta_A}{\beta_B} = 10^{\frac{\pm\Delta P_{in|dBm}}{10}} \tag{F-26}$$

where

$$\pm\Delta P_{in|dBm} = \text{wanted diode match}$$

As an example, suppose several detectors are needed to be matched to within ±0.5 dB.

$$\frac{\beta_A}{\beta_B} = 10^{\frac{+0.5}{10}} = 1.12 \; (\text{for } +0.5 \; \text{dB}) \tag{F-27}$$

$$\frac{\beta_A}{\beta_B} = 10^{\frac{-0.5}{10}} = 0.89 \; (\text{for } -0.5 \; \text{dB}) \tag{F-28}$$

Thus, for a ±0.5 dB match

$$0.89 < \frac{\beta_A}{\beta_B} < 1.12 \tag{F-29}$$

168

It has been found experimentally that matching the diodes at –20 dBm gives a match within ±0.5 dB over their full dynamic range. This greatly eases diode-matching problems.

Table F-5 offers a performance comparison of Schottky and tunnel diode detectors. With the advent of planar tunnel diodes [6], the mechanical deficiencies of the mesa structure are avoided, and operating temperatures in excess of 100°C are possible. If the limited dynamic range of tunnel diode detectors can be tolerated, they are a logical choice for logarithmic amplifier applications.

TABLE F-5. Comparison of Schottky and Tunnel Diode Detectors.

Diode characteristics	Schottky	Tunnel (see Figure F-9)	See references
Bias	50 to 300 μA	0	1, 2, 3, 4
T_{ss} B_V = 2 MHz, N_F = 3 dB	–50 to –52 dBm	–49 dBm	1
Video resistance, R_V (square law)	200 to 400Ω	100 to 120Ω	1, 4, 6
VSWR (total)	4:1	1.5:1	1, 4, 6
Frequency flatness	±1.5 dB	±0.5 dB	1, 2, 3, 4, 6
Relative rise and fall time	Moderate to long	Short	1
Temperature stability	±1 dB	±0.5 dB	1, 4, 6
Power rating	+20 dBm	+17 dBm	1
Dynamic range	–50 to +17 dBm	–49 to +5 dBm	1, 3, 4, 6
Matched units	Difficult	Easy	1, 3, 4

References

1. Hewlett-Packard. *Hot Carrier Diode Video Detectors*. (Application Note 923, no date.)

2. Aertech Industries. *Detectors. Limiter Detectors. Comb Generators*. (Catalog No 9778 no date)

3. Naval Weapons Center. *On Certain DC-Coupled Detector/Video Amplifiers*, by R. S. Hughes China Lake Calif. NWC, July 1984. (NWC TM 5279.)

4. Hughes. R.S., "Tunnel Diodes Excel as DC-Coupled Detectors" *Microwaves* June 1981 pp. 59-62.

5. Ibid.

6. Tatum, J. and K. Hinton, "Tunnel Diodes Complement High-Performance Detectors," *Microwaves and RF*, February 1985, pp. 115-124.

Appendix G

TANGENTIAL SIGNAL SENSITIVITY

Tangential signal sensitivity is a useful method of comparing receiver sensitivities [1]. Figure G-1 illustrates the basic measurement techniques [2]. The reader should consult the references and the Bibliography for a complete discussion.

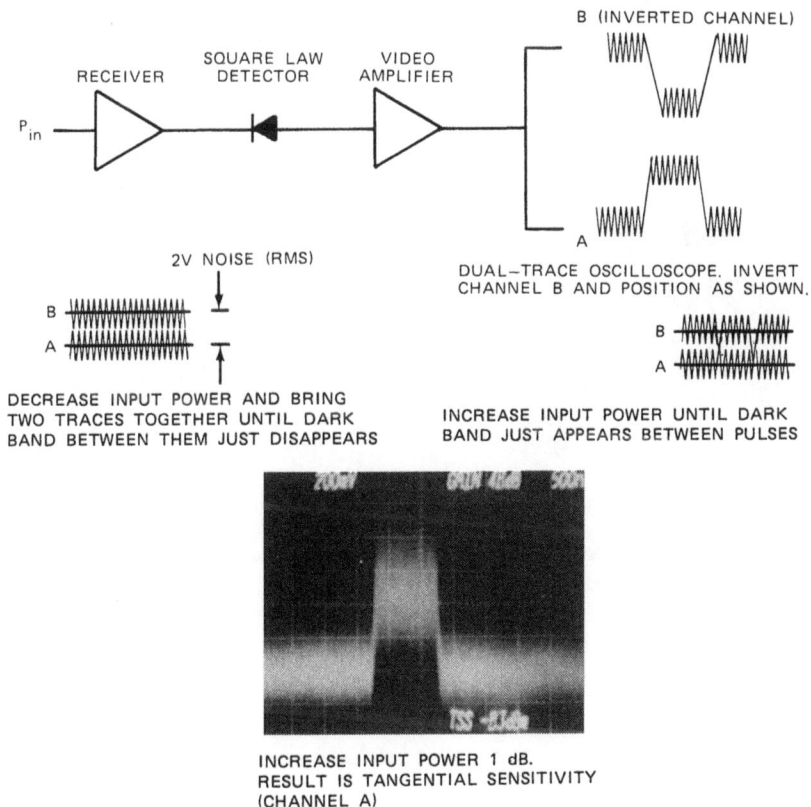

FIGURE G-1. Tangential Sensitivity Measurement.

171

References

1. Lucas, W. J., "Tangential Sensitivity of a Detector Video System With RF Preamplification," *Proc. IEEE*, Vol 113, No. 8, August, 1966, pp 1321–1330.

2. Hughes, R. S. "Practical Approach Makes TSS Measurement Simple," *Microwaves*, January 1982, pp 69–72.

Appendix H

CONTINUOUS WAVE REJECTION

It is possible to design detector-logarithmic video amplifiers, using either tunnel or Schottky diode detectors, insensitive to CW inputs of up to some -15 dBm. At the present time this is not easily possible for logarithmic IF amplifiers.

Figure H-1 illustrates the basic detector-logarithmic video amplifier discussed in Chapters 2 and 4.

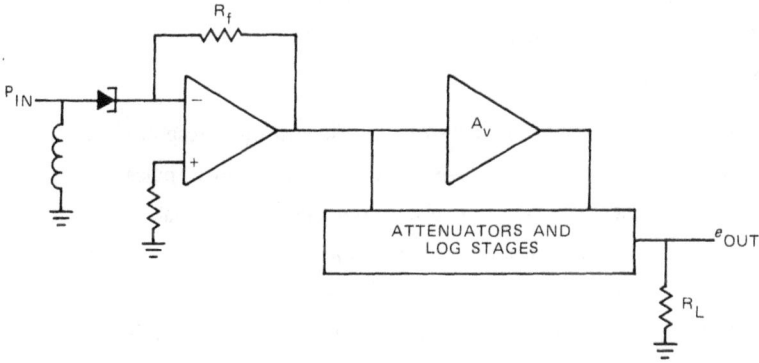

FIGURE H-1. Basic Detector-Logarithmic Video Amplifier.

With reference to Figure H-1, the input power is detected by the tunnel diode/transimpedance amplifier combination, and it drives the high-level logarithmic stages and the fixed-gain amplifier, A_v. The output of the fixed-gain amplifier drives the lower–level logarithmic stages. The equation for the detector/transimpedance ampli-fier (see Appendix F) may be given as

$$e_D = \beta R_f \left[10^{\frac{P_{in|dBm}}{10}} \right]$$ (H-1)

where

$$\beta = \text{detector short circuit current sensitivity}$$
$$P_{in|dBm} = \text{input power in dBm}$$
$$R_f = \text{feedback resistor}$$

β is fairly constant for input powers less than 25 dBm or so. However, as $P_{in|dBm}$ is increased, the detector deviates from its theoretical "square law" response, as shown in Figure H–2. It is a fairly simple matter (as shown in Chapter 2) to compensate for this deviation from square law in the logarithmic amplifier; the detector–logarithmic video amplifier output is shown in Figure H–3. Figure H–4 illustrates the detector–logarithmic video amplifier output for pulsed inputs (output normalized).

To measure the effects of CW on the logarithmic response of pulse signals, the test configuration of Figure H–5 is used. For a given pulse signal input, the CW signal can be increased and its effect on the pulse signal noted.* Figure H–6 illustrates the detector-logarithmic video amplifier output for a pulsed signal of –25 dBm and four CW magnitudes (no CW, –40, –30, and –20 dBm). Obviously, –20 dBm CW input renders the detector-logarithmic video amplifier useless for the –25 dBm pulsed signal, and a –30 dBm CW input represents an increase of some 90 mV in the pulsed signal output. Since the logarithmic slope is 43.5 mV/dB (see Figure H–2), the 90-mV pulse increase represents an effective 2.07–dB error, which may well be excessive for certain applications (i.e., monopulse direction finding). To obtain a better understanding of why Figure H–6 looks the way it does, the composite detector input power will be found.

$$P_{in} = P_{signal} + P_{CW} \text{ (watts)}$$ (H-2)

---

*For the moment f_{CW} is 50 MHz greater than f_{signal}.

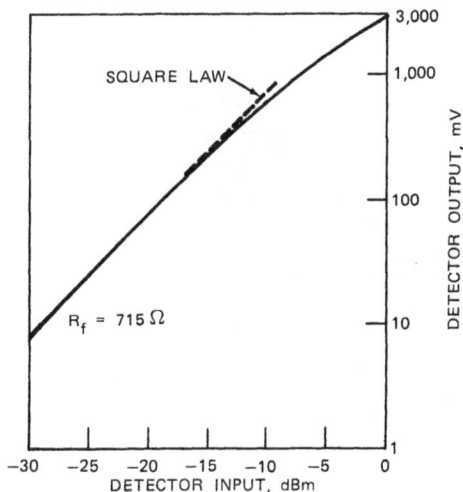

FIGURE H-2. Tunnel Diode Output Versus Input Power.

where

$$P_{signal} = 1 \times 10^{-3} \; 10^{\frac{P_{signal}|_{dBm}}{10}} \quad \text{(watts)} \tag{H-3}$$

$$P_{CW} = 1 \times 10^{-3} \; 10^{\frac{P_{CW}|_{dBm}}{10}} \quad \text{(watts)} \tag{H-4}$$

combining Equations H-2, H-3, and H-4,

$$P_{in} = 1 \times 10^{-3} \left(10^{\frac{P_{signal}|_{dBm}}{10}} + 10^{\frac{P_{CW}|_{dBm}}{10}} \right) \text{(watt)} \tag{H-5}$$

Since the detector-logarithmic video amplifier output is given in V/dB, Equation H-5 will be converted to dBm,

$$P_{in}|_{dBm} = 10 \log \left[10^{\frac{P_{signal}|_{dBm}}{10}} + 10^{\frac{P_{CW}|_{dBm}}{10}} \right] \tag{H-6}$$

To gain an understanding of the effect of Equation H-6 on the log response (Figure H-6), the logarithmic transfer function for the detector-logarithmic video amplifier of Figure H-3 may be given as (see Chapter 1)

175

$$e_{out} \cong LS(V/dB) \left[P_{in}|_{dBm} - P_{min}|_{dBm} \right] \tag{H-7}$$

where

$$LS = \text{logarithmic slope in V/dB}$$

$$P_{in}\Big|_{dBm}^{e_o=o} = \text{extrapolated input for 0 V output}$$

or, for the detector-logarithmic video amplifier response of Figure H-3

$$e_{out} \cong 43.5 \times 10^{-3} \left(46 + P_{in}|_{dBm} \right) \tag{H-8}$$

where

$$P_{in}|_{dBm} = \text{total power input as given in Equation H-6}$$

Let the following inputs be applied to the detector-logarithmic video amplifier:

$$P_{CW}|_{dBm} = -30$$

$$P_{signal}|_{dBm} = -25$$

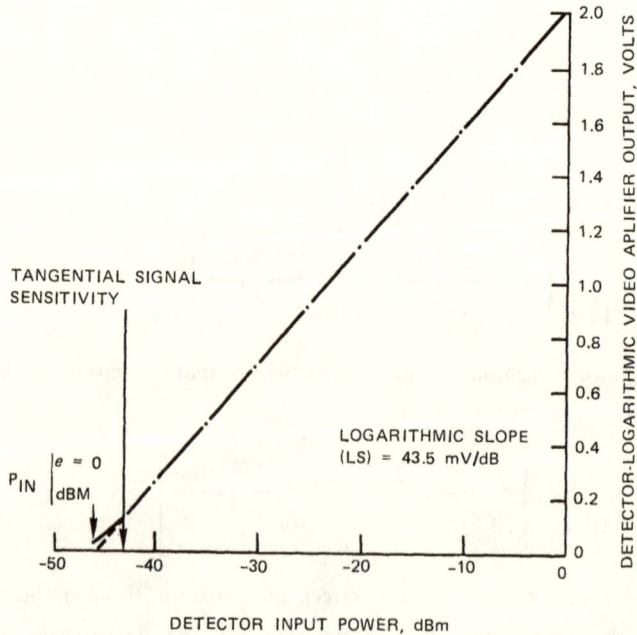

FIGURE H-3. Detector-Logarithmic Video Amplifier Output Versus Input Power.

The detector-logarithmic video amplifier output due to the CW only is (Equation H-8)

$$e_{out}|_{CW} = 43.5 \times 10^{-3} (46 - 30) = 696 \text{ mV} \tag{H-9}$$

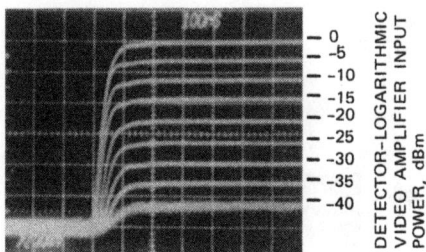

FIGURE H-4. Detector-Logarithmic Video Amplifier Pulse Characteristics.

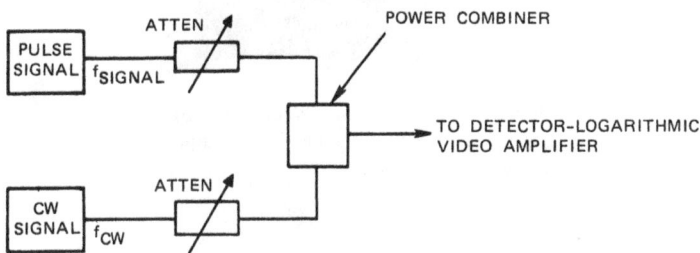

FIGURE H-5. CW Effects Test Configuration.

and this compares favorably with the measured result (see Figure H-6). The predicted detector-logarithmic video amplifier output for the composite signal will now be found. Using Equation H-6,

$$P_{in}|_{dBm} = 10 \log \left[10^{\frac{-25}{10}} + 10^{\frac{-30}{10}} \right] \tag{H-10}$$

or

$$P_{in}|_{dBm} = -23.81 \text{ dBm} \tag{H-11}$$

and from Equation H-3

$$e_{out} = 43.5 \times 10^{-3} (46 - 23.81) = 965 \text{ mV} \tag{H-12}$$

177

which also compares favorably with the measured result of Figure H–6.

Thus, the strong dependence of CW on the pulse amplitude, Figure H–6, is to be expected. All is not lost, however. Combining Equations H–6 and H–1, the detector output as a function of $P_{signal}|_{dBm}$ and $P_{CW}|_{dBm}$ is given as

$$e_D = \beta R_f \left[10^{\frac{P_{signal}|_{dBm}}{10}} + 10^{\frac{P_{CW}|_{dBm}}{10}} \right] \tag{H-13}$$

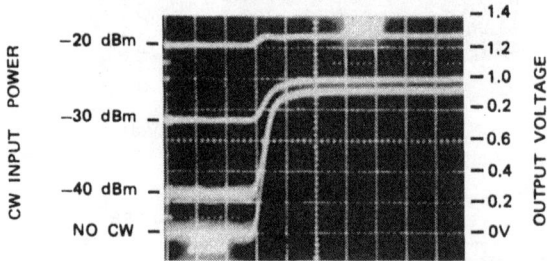

FIGURE H-6. Effect of CW on Detector-
Logarithmic Video Amplifier Pulse Output
(P_{signal} = -25 dBm).

If the second term in brackets can be driven to zero, the detector output becomes that of the signal, Equation H–1. The detector/transimpedance amplifier was modified as shown in Figure H–7. V_{CW} can be varied to drive e_{out} to zero (in the absence of P_{signal}) for a given CW input, thus leaving e_{out} to be a function of $P_{signal}|_{dBm}$ alone (see Chapter 4). Figure H–8 illustrates the detector/transimpedance amplifier signal output as a function of CW inputs. The output was driven to zero, by way of V_{CW}, for each CW input. As can be seen, the signal output is insensitive to CW for CW levels less than –15 dBm or so. This –15-dBm signal level is the point where the detector starts its deviation from square law (see Figure H–2). Another way to look at this deviation is with the aid of Equation H–13. The detector β is input power–sensitive (see Appendix F), and, for a total input power greater than 20 dBm, β (thus e_d) is dependent on the CW input power.

178

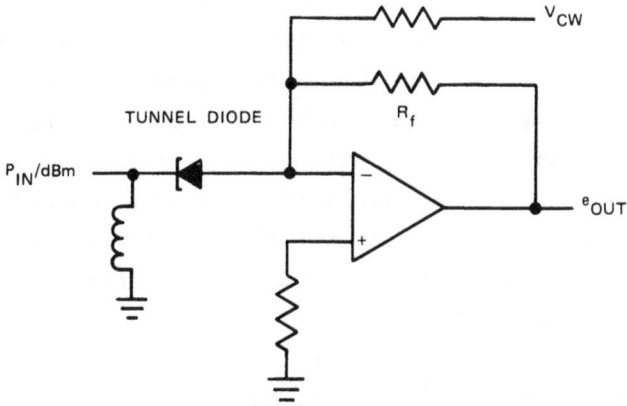

FIGURE H-7. CW Offset Control.

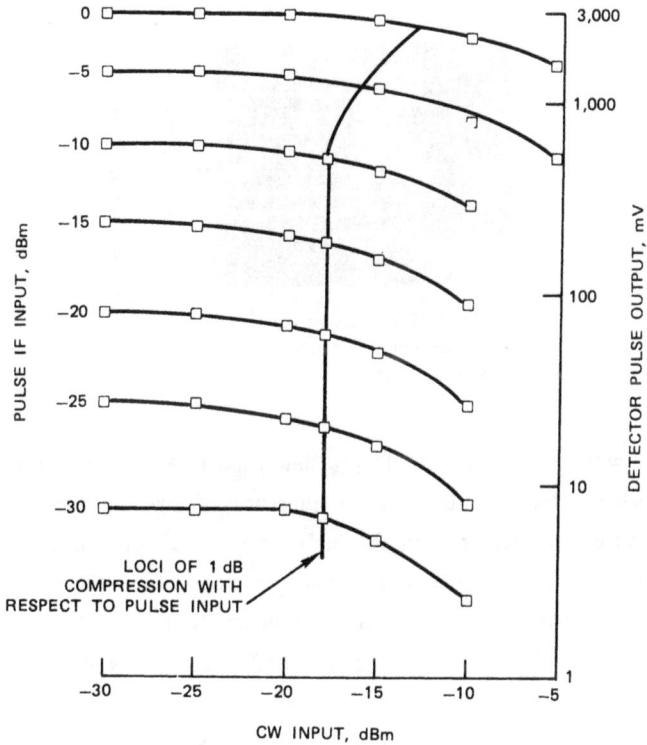

FIGURE H-8. Detector Pulse Output in Presence of CW,
CW Output Driven to Zero - Figure H-7.

179

Thus, we have a method of minimizing the effect of CW on the logarithmic transfer function for pulsed signals: Drive the detector/amplifier output to zero for CW inputs (in the absence of pulse signals), thus ensuring the logarithmic amplifier input is due to the signal input power only. This technique is valid for CW inputs less than -15 dBm or so (deviation from square law in the detector). Figure H-9 illustrates the detector–logarithmic video amplifier pulse characteristics for a -15 dBm CW input (nulled to zero as discussed). Comparing Figure H-9 with Figure H-4 (no CW input) confirms the theory presented. CW "stripping" before logarithmic amplification will give (for CW inputs less than -15 dBm or so) the same pulse logarithmic signal response as if no CW were present. For CW inputs above -15 dBm, the logarithmic response will deviate from ideal (see Figure H-8); however, CW stripping will still give far superior results than nonstripping (see Figure H-6).

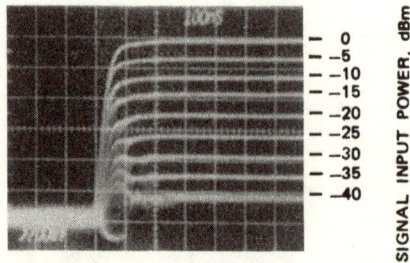

FIGURE H-9. Detector-Logarithmic Video
Amplifier Pulse Characteristics
(CW = -15 dBm).

A brief word on the effect of f_{CW} and f_{signal}. As the detector is a nonlinear device, it behaves like a mixer in the presence of two signals. If the two input frequencies differ by greater than 25 MHz or so, the detector/amplifier filters out this frequency. As the two frequencies come closer together, the detector–logarithmic output will modulate at the difference frequency (Δf), and the modulation amplitude will be dependent upon the power difference between the signal and CW inputs. Figure H-10 illustrates the effect of Δf on the detector–logarithmic video amplifier output. The signal input is -25 dBm with a CW input of -30 dBm (nulled at the detector/amplifier output, as discussed). The difference frequency ($\Delta f = f_{signal} - f_{CW}$) was varied, and, as can be readily seen, the signal modulation

180

on the detector-logarithmic video amplifier output is excessive as Δf goes toward zero. The peak-to-peak modulation in dB (referred to as the detector input) may be found by dividing the peak-to-peak modulation voltage by the logarithmic slope (LS) (see Figure H-3) and is listed in Figure H-10.

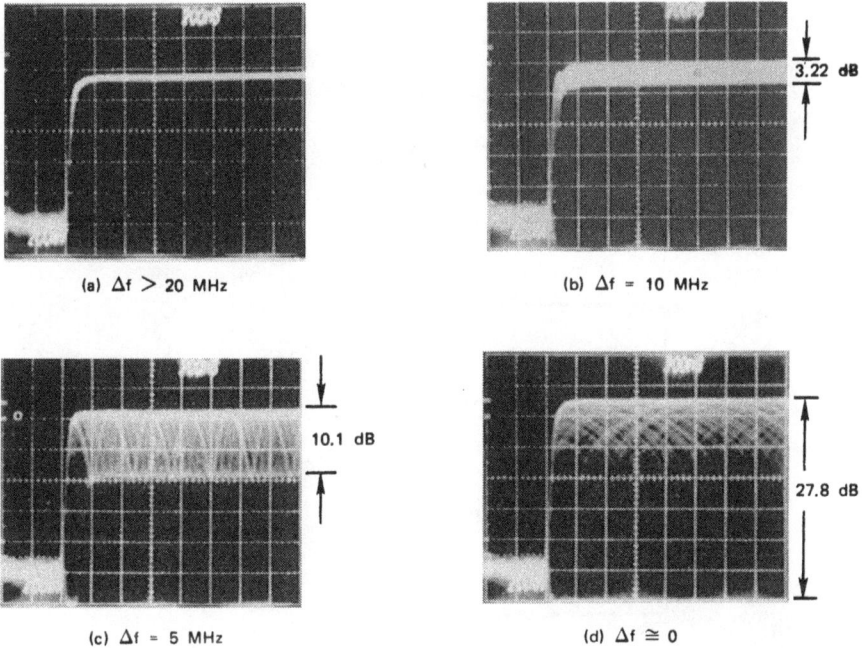

(a) Δf > 20 MHz

(b) Δf = 10 MHz

(c) Δf = 5 MHz

(d) Δf ≅ 0

FIGURE H-10. Effect of Δ on Detector-Logarithmic Video Amplifier Output Modulation (P_{signal} = -25 dBm, P_{CW} = -30 dBm [nulled at detector/amplifier output]).

Schottky diodes have simpler inherent CW rejection as shown in Figure H-11.

Logarithmic IF amplifiers, at present, do not have CW rejection capability. Figure H-12 illustrates the pulse response for a logarithmic IF in the presence of CW. The test circuit of Figure H-5 was used, and the pulse input is a constant -25 dBm. The CW input is raised from (a) no CW, (b) -50 dBm, (c) -40 dBm,

(d) −30 dBm, (e) −20 dBm, and (f) −10 dBm. The frequency difference between f_{pulse} and f_{CW} is 50 MHz. A practical equation for the maximum CW present before pulse interference is

$$P\bigg|_{CW}^{max} \cong \text{Signal input (in dBm)} -3 \qquad (\text{H}-14)$$

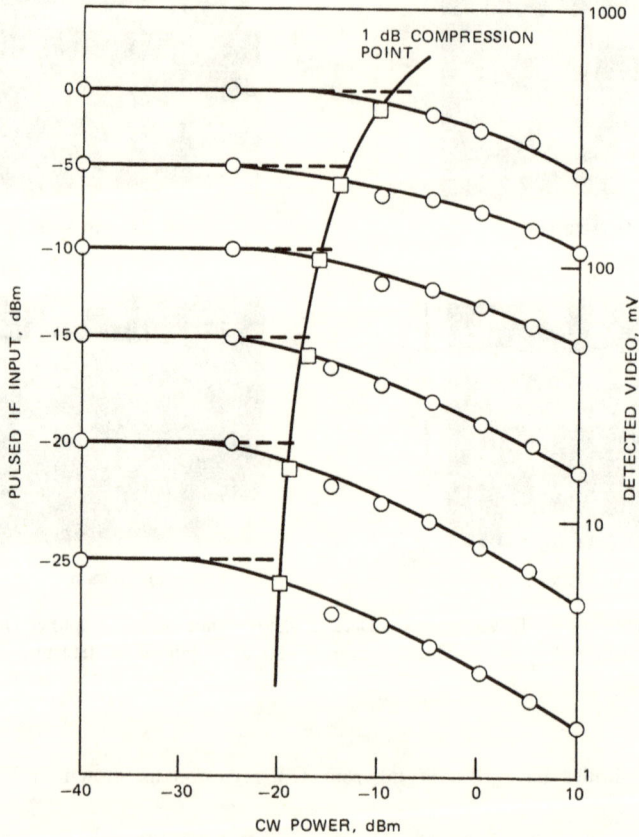

FIGURE H-11. CW Rejection for a Schottkey Diode Detector.

The true logarithmic IF amplifier, discussed in Chapter 3, will be equally CW sensitive as the CW is being logged with the pulse prior to detection.

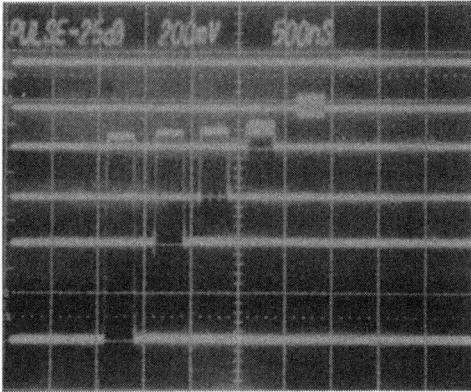

(f) CW = 10 dBm

(e) CW = -20 dBm

(d) CW = -30 dBm

(c) CW = -40 dBm

(b) CW = -50 dBm

(a) NO CW

FIGURE H-12. CW Rejection for Logarithmic IF
Amplifier (Pulse Input = -25 dBm).

183

NOMENCLATURE

AC   Alternating current

AGC   Automatic gain control

A_n Linear video stage A_n

A_o Amplifier low frequency open loop gain

$A_{V(A)}$ Voltage gain

A_Δ Voltage gain for differencing amplifier

A_Σ Gain of summing amplifier

B   Differential amplifier bias constant ($I_T R_C/2$)

BW   3–dB bandwidth for a linear amplifier

$\mathcal{C}_L$ Center of logarithmic action for differential amplifier

CW   Continuous wave

dB   Decibels

dBm   Decibel referred to 1 milliwatt

DC   Direct current

DF   Direction finder

DLVA   Detector–logarithmic video amplifier

e_d Detector output voltage

e_{in} Input voltage

$e_{in}\big|_{dB}$ Input voltage in dB

$e_{in}\Big|_{t}^{dB}$ Linear–logarithmic transition input voltage

$e_{in}\Big|_{dB}^{e_o=0}$ Logarithmic amplifier input voltage for zero output voltage

$e_{in}\Big|_{dB}^{max}$ Maximum input for logarithmic action

$e_{in}\Big|_{max}^{+}$ Maximum positive going input for $e_o\big|_{max}$

e_L Limited voltage

e_o Output voltage

184

$e_o\Big|_t^{dB}$	Output voltage for $e_{in}\dfrac{dB}{t}$
$e_o\Big|_{max}^{-}$	Maximum negative going output voltage
$e_{out}\Big|_{P_{in}}^{actual}$	Actual tunnel diode–linear amplifier output for a given input power
EW	Electronic warfare
$e_x|dB$	Voltage x, dB
$G(\theta)$	Antenna gain as a function of angle (θ)
GHz	Gigahertz
$h_{FE}(\beta)$	Transistor short circuit current sensitivity
I	Input intensity at antenna face
I_{BIAS}	Detector bias current
I_c	Collector current
ID	Diode current
IDR	Instantaneous input dynamic range, dB
I_E	Emitter current
IF	Intermediate frequency
I_S	Transistor reverse saturation current
I_T	Differential amplifier bias current $(I_{E_1} + I_{E_2})$
K	Boltzmann's constant or detector constant
K_a	Exp $\Delta V/V_T$
K_b	$1/K_a$
K_1	Logarithmic amplifier constant
K_2	Logarithmic amplifier constant
K_3	$K_1 \log K_2$
L_n	Logarithmic stage n
LS	Logarithmic slope, volt/dB
MHz	Megahertz
mV	Millivolts

N(n)	Number of logarithmic stages
nsec	Nanosecond
op-amp	Operational amplifier
pA	Pico amp
$P_{in}\big|_t^{dBm}$	Input power for linear–logarithmic transition
$P_{in}\big|_{dBm}^{e_o=0}$	Input power for $e_o = 0$
PRF	Pulse repetition frequency
PW	Pulse width
q	Electron charge
R	Ratio of logarithmic amplifier inputs
R_C	DC collector resistor
r_c	Effective AC collector resistance
R_E	Emitter resistance
R_e	Unbypassed emitter resistance
r_e'	Dynamic emitter resistance
RF	Radio frequency
R_f	Operational amplifier feedback resistance
R_L	Logarithmic amplifier load resistance or detector load resistance
R_x	Tunnel diode detector VSWR optimization resistance
S/N	Signal-to-noise ratio
T	Temperature, deg Kelvin
TDR	Total input dynamic range, dB
T_{ss}	Tangential signal sensitivity
V	Volts
V_B	Base voltage
V_C	Collector voltage
V_E	Emitter voltage

VSWR	Voltage standing wave ratio
V_T	KT/q
Z	Tunnel diode normalization factor
β	Detector short circuit current sensitivity
Δ	Difference of two signals
ΔV_{BE}	Transistor base — emitter voltage match
ΔVD	Diode voltage match
ΔID	Diode current match
ΔT	Temperature change
ϵ_{dB}	Logarithmic error with respect to ideal–logarithmic conformity
θ	Angle of two signals
μsec	Microseconds
μV	Microvolt
Σ	Sum
$\tau_r\big|_{10\%}^{90\%}$	Rise time definition for a linear receiver
ϕ	V_{BE}

BIBLIOGRAPHY

This bibliography lists the subjects in chronolgical order. My reason for this is to pay tribute to those whose works we build on.

LOGARITHMIC AMPLIFICATION

Corney, J. "A Simple Logarithmic Receiver," in *Proceedings of IRE*, Vol. 39, N0.7 (July 1951), pp. 807-813.

Chambers, T. H. and I. H. Page. "A High Accuracy Logarithmic Receiver," in *Proceedings of the IRE*, August 1954, pp. 1307-1314.

Rozenstein, S. "Design of a Logarithmic Receiver," in *Proceedings of the IRE*, Vol. 102, Part B (January 1955), pp. 69-74.

Kihn, H. and W. E. Barnette. "A Linear-Logarithmic Amplifier for Ultra-Short Pulses," *RCA Review*, March 1957, pp. 95-135.

Stanford University, Stanford Electronics Laboratories. *Some Log Video Amplifier Analysis and Design Techniques*, by J. C. de Broekert. Stanford, Calif., SU-SEL, 5 April 1957. (Technical Report No. 152-1.)

Mathams, R. F. "A Voltage Operated Logarithmic Amplifier," in *Electronic Engineering*, August 1959, pp. 463-465.

Solms, S. J. "Logarithmic Amplifier Design," in *IRE Transactions on Instrumentation*, December 1959, pp. 91-96.

Stanford University, Stanford Electronics Laboratories. *A Solid State Logarithmic Video Amplifier*, by G. S. Bahrs and W. D. Hindson. Stanford, Calif., SU-SEL, 22 April 1960. (Technical Report No. 755-2.)

- - - - -. *A Wideband Limiting—Summation Logarithmic Video Amplifier Design*, by W. R. Kincheloe. Stanford, Calif., SU-SEL, 6 June 1960. (Technical Report No. 560-1.)

Stanford University, Stanford Electronics Laboratories. *A Solid State Analogue Video Multiplier (Circuit Technique)*, by W. H. Huntley Jr., E. Tammaru, and F. Behr. Stanford, Calif., SU-SEL, July 1962. (Technical Report No. 806-2.)

Alcock, R. N. "A Wide Band Transistor Logarithmic Amplifier at 45 Mc/S," *Electronic Engineering*, July 1962, pp. 444-449.

U.S. Naval Research Laboratory. *A Solid State Logarithmic Video Amplifier for Pulse Applications*, by R. L. Slice. Washington, D.C., 11 August 1964. (Laboratory Report No. 6147, publication UNCLASSIFIED.)

Stanford University, Stanford Electronics Laboratories. *A Method of Obtaining a Log Video Amplifier With a 100 dB Dynamic Range*, by R. M. Kochis. Stanford, Calif., SU-SEL, 18 November 1964. (Technical Note No. 1969-A.)

Pearlman, A. R. "Logarithmic and Hyperlogarithmic Signal Amplifiers," in *EDN*, October 1965, pp. 85-92.

Syracuse University, Syracuse University Research Corporation. *A High Performance Log Video Amplifier*, by P. E. Harris. Syracuse, New York. October 1965. (Report No. DSL R-136.)

Stanford University, Stanford Electronics Laboratories. *Design for a High-Duty-Cycle, Bipolar, Logarithmic Video Amplifier*, by R. L. McMaster. Stanford, Calif., SU-SEL, October 1965. (Technical Report No. 1972-2.)

Admiralty Surface Weapons Establishment. *Phase Stability of the Bi-Gain Stage Type of True I. F. Logarithmic Amplifier*, by A. Woroncow and J. Croney. Portsmouth, England, 14 October 1965. (ASWE Lab Note XRA-65-7, publication UNCLASSIFIED.)

Volkov, V. M. "Transistorized Logarithmic Amplifiers," in *Foreign Technology Division Translation FTD-MT-24-118-67 of the Russian 1965 Book.*

Deighton, M. O., E. A. Sagle, and C. S. Kestin. "Analogue Pulse Arithmetic Using Logarithmic p-n Junctions," in *Proceedings of the IEE*, Vol. 113, No. 2 (February 1966), pp. 225-233.

Defence Research Telecommunications Establishment. *A Logarithmic IF Amplifier for CW or Pulse Type Signals*, by F. R. Cross and D. C. Barnes. Ottawa, Ontario, February 1966. (Report No. DRTE-1156, publication UNCLASSIFIED.)

Platzer, G. E. Jr. "Using Transistor Circuits to Multiply and Divide," in *Electronics*, 4 April 1966, pp. 109-115.

U.S. Naval Ordnance Test Station. *A Transistorized Logarithmic Video Amplifier,* by R. S. Hughes. China Lake, Calif., NOTS, May 1966. (NOTS TP 4099, publication UNCLASSIFIED.)

Gay, M. J. "Log IF Strips Use Cascaded ICS," *Electronic Design,* 19 July 1966, pp. 56–59.

Hirsch, R. "Four Pitfalls in Using Log Amplifiers," in *EEE,* August 1966, pp. 122–123.

Woroncow, A. and J. Croney. "A True IF Logarithmic Amplifier Using Twin–Gain Stages," *The Radio and Electronic Engineer,* September 1966, pp. 149-155.

Brown, A. J. "Waveshape Distortion in Log Receivers," in *EEE,* September 1966, pp. 96-138.

Rubin, S. N. "A Wide–Band UHF Logarithmic Amplifier," *IEEE Journal of Solid-State Circuits,* Vol. SC-1, No. 2 (December 1966), pp. 74-81.

U.S. Naval Ordnance Test Station. *Logarithmic Video Amplification Using Sequential Current Summation,* by R. S. Hughes. China Lake, Calif., NOTS, June 1967. (NOTS TP 4375, publication UNCLASSIFIED.)

Hughes, R. S. "A Dual-Polarity Logarithmic Video/IF Amplifier Technique," in *Paper Presented at IEEE International Conference on Communications,* 12–14 June 1967.

Syracuse University, Syracuse University Research Corporation. *A High Duty Cycle, High Performance Log Video Amplifier,* by P. E. Harris. Syracuse, New York, 15 August 1967. (Report No. DSD TM-181.)

Hughes, R. S. "Semiconductor Variable Gain and Logarithmic Video Amplifiers," *The Continuing Education Institute, Inc.,* Phoenix, Arizona, 1967.

Pennsylvania State University, Pennsylvania State University Department of Electrical Engineering. *A High Precision Logarithmic Amplifier Using Sampling Techniques,* by L. R. Simmering. University Park, Pa., 1 December 1967. (Scientific Report No. 313.)

Johnson, W. "Logarithmic Compression of Radio Frequencies of Various Signal Waveforms," *The Aerospace Corporation,* Los Angeles, Calif., 31 July 1969 (Report No. TR-0066(5230-46)-4.)

Hughes, R. S. "New Log Amp Cascades to Desired Range," *Electronic Design,* 25 October 1969, pp. 86-89.

Borlase, W. and E. David. "Design of Temperature-Compensated Log Circuits Employing Transistors and Operational Amplifiers," *Analogue Devices*, Norwood, Mass., October 1969. (Report No. E020a-10-9/69.)

Dobkin, R. C. "Logarithmic Converters," *IEEE Spectrum*, November 1969, pp. 69-72.

Hughes, R. S. "Make Very Wide-Range Log Amps Easily," *Electronic Design*, 11 October 1970, pp. 76-78.

Lansdowne, K. and A. J. Kelly. "Microwave Logarithmic Amplifiers Using Hybrid-Integrated Circuits," *IEEE International Solid-State Circuits Conference*, 1971, pp. 94, 95.

Hughes, R. S. *Logarithmic Video Amplifiers*, Artech House, Inc., Dedham, Mass., 1971.

Smith, G. 3rd. "Try a Piecewise-Linear Approach to the Design of Wide-Range Log Amplifiers," *Electronic Design*, 6 January 1972, pp. 66-70.

Clifford, D. "Approximating True Log Output at High Frequencies," *Electronics*, 31 January 1972, pp. 70-72.

Naval Weapons Center. *Logarithmic Video Amplifiers*, by R. S. Hughes. China Lake, Calif., NWC, March 1972. (NWC TP 5333, publication UNCLASSIFIED.)

Morgan, D. R. "Get the Most Out of Log Amplifiers by Understanding the Error Sources," in *EDN*, 20 January 1973, pp. 52-55.

Ehrsam, B. "Transistor Logarithmic Conversion Using an Integrated Operational Amplifier," *Motorola Semiconductor Report AN-261*, (No date).

Niv, G. "Get Wide Dynamic Range in Log Amps," *Electronic Design*, 15 February 1973, pp. 60-62.

Loesch, B. "A UHF Logarithmic IF Amplifier," *IEEE Transactions on Aerospace and Electronic Systems*, Vol. AEJ-9, No. 5 (September 1973), pp. 660-664.

Lipsky, S. E. "Log Amps Improve Wideband Direction Finding," *Microwaves*, May 1973, pp. 58-65.

Risley, A. R. "Designing Guide to: Logarithmic Amplifiers," in *EDN*, 5 August 1973, pp. 42-51.

Sheingold, D. and J. Cadogan. "A Guide to Specifying and Testing Logarithmic Devices," in *EDN*, 20 September 1973, pp. 54-58.

Sheingold, D. and F. Pouliot. "The Hows and Whys to Logarithmic Amps," *Electronic Design,* 1 February 1974, pp. 52–59.

Helfrick, A. "Build High–Gain, Wide–Range Log Amps," *Electronic Design,* 15 March 1974, pp. 116–118.

Lamagna, J. "Design Compact Log Amps," *Microwaves,* April 1974, pp. 58–61.

Sareen, S. "Direct–Coupled Detector Log Amplifier for Crystal Video Signal Processing," *Microwave System News,* April/May 1975, pp. 85–89.

Barber, W. L. and E. R. Brown. "A True Logarithmic Amplifier for Radar IF Applications," *IEEE Journal of Solid–State Circuits,* Vol. SC–15, No. 3 (June 1980), pp. 291–295.

Sheade, M. "DLVAS Find Application in ESM Systems," *MSN,* August 1984, pp. 47–56.

Plessy Company. *"Broadband Amplifiers Applications,"* Irvine, Calif., September 1984.

Larsen, D. "Logarithmic Converter Handles Wide Dynamic Ranges," *Electronic Products,* 1 March 1985, pp. 69–75.

Potson, D. and R. S. Hughes. "DC Coupled Video Log Amp Processes 10 nsec Pulses," *Microwave and RF,* Part 1, April 1985, pp. 85–90, 150. Part 2, May 1985, pp. 75–78, 278–280.

Counts, L., C. Kitchen, and S. Sherman. "One–Chip 'Slide Rule' with Logs, Antilogs for Real–Time Processing," *Electronic Design,* 2 May 1985, pp. 121–128.

DETECTORS

The following is a listing of various detector–oriented works the author has found most helpful. The reader is also referred to the catalogs of the various detector manufacturers.

Mouw, R. B. and F. M. Schumacher. "Tunnel Diode Detectors," *Microwave Journal,* January 1966, pp. 27–36.

Tenenholtz, R. "The Video Detector," in *A Technical Discussion,* Sage Laboratories, Inc., Natick, Mass., 1968.

Hewlett Packard. *Hot Carrier Video Detectors,* Palo Alto, Calif., (No Date). (Application Note 923.)

Turner, R. J. "Schottky Diode Pair Makes an RF Detector Stable," *Electronics,* 2 May 1974, pp. 94, 95.

Hewlett Packard. *Temperature Dependence of Schottky Detector Voltage Sensitivity,* Palo Alto, Calif., October 1975. (Application Note 956-6.)

-----. *Dynamic Range Extension of Schottky Detectors,* Palo Alto, Calif., October 1975. (Application Note 956-5.)

Hughes, R. S. "Tunnel Diodes Excel as DC-Coupled Detectors," *Microwaves,* June 1981, pp. 59-62.

Hewlett Packard. *Square Law and Linear Detection.* Palo Alto, Calif., September 1981. (Application Note 986.)

Tatum, J. and K. Hinton. "Tunnel Diodes Complement High-Performance Detectors," *Microwaves and RF,* February 1985, pp. 115-124.

AUTOMATIC GAIN CONTROL

Automatic gain control and logarithmic application often go hand in hand, so a short listing of automatic gain control works is warranted.

Oliver, B. M. "Automatic Volume Control as a Feedback Problem," *Proceedings of IRE,* April 1948, pp. 466-473.

Field, J. C. G. *The Design of Automatic-Gain-Control Systems For Auto-Tracking Radar Receivers,* The Institute of Electrical Engineers, Monograph No. 258R, October 1957, pp. 93-108.

Victor, W. K. and H. H. Brockman. "The Application of Linear Servo Theory to the Design of AGC Loops," *Proceedings of IEEE,* February 1960, pp. 234-238.

Rheinfelder, W. A. "Design Automatic Gain Control Systems," in *EEE,* Part 1, December 1964, pp. 43-47. Part 2, January 1985, pp. 53-57.

Stanford University, Stanford Electronics Laboratories. *Automatic Gain Control Theory for Pulsed and Continuous Signals,* by E. W. Senior. Stanford, Calif., SU-SEL, June 1967. (Technical Report No. 1606-2.)

Ossott, A. "Design of A Solid State IF AGC System for Pulsed Carrier Microwave Receivers," *Microwave Journal,* July 1967, pp. 43–48.

Naval Weapons Center. *Automatic Gain Control: A Practical Approach to its Analysis and Design,* by R. S. Hughes. China Lake, Calif., NWC, August 1977. (NWC TP 5948, publication UNCLASSIFIED.)

Hughes, R. S. "Design Automatic Gain Control Loops the Easy Way," *EDN,* 5 October 1978, pp. 123–178.

Porter, J. "AGC Loop Design Using Control System Theory," *RF Design,* June 1980, pp. 27–32.

SENSITIVITY

The designer and user of logarithmic amplifiers ultimately must face the sensitivity problem. It is hoped that this short bibliography proves useful.

Stanford University, Stanford Electronics Laboratories. *Characteristics of Crystal-Video Receivers Employing RF Preamplification,* by W. E. Ayer. Stanford, Calif., SU–SEL, 20 September 1956. (Technical Report No. 150-3.)

– – – – –. *Noise Figures, Noise Temperatures, and System Sensitivity,* by P. H. Enslow. Stanford, Calif., SU–SEL, 10 July 1960. (Technical Report No. 516-2.)

Klipper, H. "Sensitivity of Crystal Video Receiver With RF Preamplification," *Microwave Journal,* Vol. 8, 1965, pp. 85–92.

Lucas, W. J. "Tangential Sensitivity of a Detector Video System With RF Preamplification," *Proceedings of IEEE,* Vol. 113, No. 8 (August 1966), pp. 1321–1330.

Motchenbacher, C. D. and F. C. Fitchen. *Low-Noise Electronic Design.* John Wiley and Sons, New York, New York, 1973.

Hewlett Packard. *The Criterion for the Tangential Sensitivity Measurement.* Palo Alto, Calif., October 1973. (Application Note 956-1.)

King, D. "Determining Optimum Gain for Maximum Sensitivity," *Microwaves,* September 1979, pp. 57, 58.

Tsui, J. "Tangential Sensitivity of EW Receiver," *Microwave Journal,* October 1981, pp. 99–102.

Hughes, R. S. "Practical Approach Makes T_{SS} Measurement Simple," *Microwaves,* January 1982, pp. 69-72.

Tsui, J. and R. Shaw. "Sensitivity of EW Receivers," *Microwave Journal,* November 1982, pp. 115-120.

Tsui, J. "False Alarm Measurements on Receivers," *Microwave Journal,* September 1984, pp. 213-217.

Naval Weapons Center. Determining Maximum Sensitivity and Optimum Maximum Gain for Detector-Video Amplifier, With RF Preamplification, by Richard Smith Hughes, China Lake, Calif., March 1985. (NWC TM 5337).

www.ingramcontent.com/pod-product-compliance
Lightning Source LLC
Chambersburg PA
CBHW021431180326
41458CB00001B/215